MONTH BY MONTH
in the loft

FRANK HALL

The Racing Pigeon Publishing Co Ltd

©The Racing Pigeon Publishing Co Ltd 1992

All rights reserved. No part of this publication may be reproduced, stored in a retrieval system or transmitted in any form or by any means, electronic, mechanical, photocopying, recording or otherwise without the prior permission of The Racing Pigeon Publ Co Ltd.

This book is sold subject to the condition that it shall not, by way of trade or otherwise, be lent, re-sold, hired out or otherwise circulated without the publisher's prior consent in any form of binding or cover other than that in which it is published and without a similar condition including this condition being imposed on any subsequent purchaser.

Text set in 10/11 Century Schoolbook by
RP Typesetters Ltd, for
THE RACING PIGEON PUBLISHING CO LTD,
Unit 13, 21 Wren Street,
London WC1X 0HF,
United Kingdom

Cover picture by Colin Osman

Printed and bound in England
by Hollen Street Press Ltd, Slough.

CONTENTS

Introduction   5

JANUARY
J W Barker — Paramyxovirus — Coccidiosis — Ground Nut Oil — Symptoms and Similarities   7

FEBRUARY
Loft Conversions — Early Breeding — Colour Breeding — Veterans — Fred Marriott — Old Billy and Forlorn Hope   23

MARCH
Older Breeders — Barley — Spots in the Throat — Memory Lane — Canker — Darwin and the Survival of the Fittest   41

APRIL
Buying Pigeons — Winning Hens — The Kestrel — Arthur Sheppard — Colour Breeding — Dordins — Loft Design   55

MAY
Form and Condition — Educating Babies — Early Widowhood — Pellets — Caring for Breeders — Going Light — Worms   75

JUNE
The Greater Distances — Dealing with Failure — Young Bird Curriculum — Judging Breeders — International Barcelona Hens   93

JULY
Outstanding Beeders — Marathons — Selection — Racing Ability — Peppiatt's 'White Hope' — Young Birds and Novices — Heatwaves   107

AUGUST
YB Losses and Fitness — Late Breds — Transporters — Early YB Training — Nest Boxes — Separation — Champion Major   127

SEPTEMBER
YB Education — The Litchfield Crate — Clashing — Transporter Crates — Holbard & Son — Filling Eggs — The 1986 Rome Race   143

OCTOBER
The Moult — Separating — Alf Baker — Pure strains — Stichelbauts — Huyskens-Van Riel — Pigeon Management   163

NOVEMBER
Outer Flights — Bloomfield and Fabry — Logan and Old 86 — Busschaerts — Early Breeding Preparations — Show Racers   183

DECEMBER
The 1913 Rome Race — Thorougoods — Old Timers — Rene Boizard — Minerals — Conditioning Cake — Good Diet and Feeding — Winter Food and Water   201

INTRODUCTION

BY COLIN OSMAN

It is an unenviable task to write the introduction to this book because how does one write about a legend in his own lifetime? In the book Frank writes about his early days so it is easy to leave the biographical details to him but how does one explain the tremendous popularity of his monthly column that has run for so many years?

Older readers will not need to have it explained but for the benefit of younger fanciers I will try and explain. Frank Hall is a butcher by trade and before he retired a very good one too. Now he has passed on his skills to his son but his passion always was racing pigeons. He lived pigeons and even if he never won the National or the Combine there were 101 good reasons and no one doubts that if life's cards had fallen a different way he would have been in the star-spangled galaxy. That is not to say he has not had his share of success and when conditions permitted in the past he won many a prize.

Many years ago my father persuaded Frank to pick up the hammer and become an auctioneer. I was at that auction in the now demolished CIU headquarters in Clerkenwell Road and no one would have guessed this was his very first auction. He went on to become the top auctioneer of this period writing his auction lists with such care and meticulousness that they became the standard all others aimed at. In this he was helped by his incredible memory, something that mere mortals like myself were deeply envious of. He could not only memorise a pedigree at first reading but it was so assimilated that it was ready for use with the next pedigree. As a result his sale lists did not have errors and were genuinely informative.

As a natural corollary of his command of the facts he built up a respect unequalled in the trade and was completely trusted by buyer and seller. His sales were not only unforgettable theatrical events but object lessons on pedigress and performances. He commanded a respect the like of which was never known before or since.

This ability had of course always been there but was known only to a few friends not as now to the whole Fancy. Over the years I have seen many of the pedigrees he has prepared and they are calligraphic works of art. More importantly they were absolutely correct in every detail and totally reliable.

When he started his monthly articles for The Racing Pigeon many years ago, the immediate thing that impressed was this memory for detail. When he told the histories of the great fanciers of the past you knew immediately they were accurate. But there was more. Much more. This was not a dry recital of facts but a deeply intuitive assessment of the fancier as a fancier. Because Frank knew and understood every aspect of pigeon racing he could in his writings bring out the true heart of a fancier. He could make them come alive. He could make you the reader understand how these fanciers achieved greatness. It is a rare gift and no one does it better than Frank.

Added to that is an enquiring mind and a very keen sense of observation. If he investigates a topic, as in these pages, he comes with no preconceived ideas but with an open mind and his discoveries he shares with his readers. He is one of the greatest writers in pigeon racing and that it is why I feel privileged to write this introduction.

Now to a few practicalities. If Frank had a fault, it was taking on too much work and the practical result was that Month by Month was slipped through my letterbox at 4am when Frank was on his way to Smithfield Market. No harm in that except that it was always the last week of the month so that the article would appear at the end of the month rather than the beginning. This should be borne in mind when reading the monthly chapters.

There is also the point that the monthly articles often produced a lively response the following month or even later. In editing the monthly articles I have tried to combine not only what Frank said relevant to the month in question over several years but also to combine together several months on one topic to make smoother reading. If there are faults in the editing they are mine not Frank's. Each chapter has been taken from several years and blended together. Where necessary I have judiciously made minor amendments to make the blending easier to follow. One or two subjects, for reasons of space, have been omitted but hopefully they will appear in a sequel. As an editor I should be inconspicuous and I shall be well pleased if I have helped in the re-creation of some of the best writing of Frank Hall in more permanent form.

JANUARY

J W Barker — Paramyxovirus — Coccidiosis — Ground Nut Oil — Symptoms and Similarities

It was in December 1956 staying with my very dear friends the late Mr & Mrs Ashman at Old Hill, Staffs, that I wrote the first Month by Month for January. The snow was "deep and crisp and even" outside and I was warm and cosy inside Harry and Nancy's delightful house now alas gone in the wake of developers. That delightful memory of my very first of several visits to the unique house in Cherry Orchard will remain with me until the end of my days. As indeed it doubtless will be for Harry and Nancy's son and heir Robert Ashman now of West Hagley, better known as "Bob" Ashman for his many successes with his Widowhood team of outstanding cocks, although ironically his greatest ever performer to date is that wonder hen, Dail's Delight. She won a number of outstanding prizes not least among these being 1st Sect Midlands Nat Angouleme in 1982, not raced in 1983, and won 1st Open and 1st Sect Midlands Nat Angouleme 1984 and that same year prior to Angouleme this wonderful hen won 68th Sect, 166th Open National FC Nantes and also 2nd Sect, 12th Open Midlands Nat Nantes. In 1982 this same hen, named by Robert after his wife, won 3rd Sect J, 171st Open Nantes Nat, and 15th Sect, 18th Open Midlands Nat Nantes. Other main wins for Dail's Delight included 1st Club Nantes and 1st Hare & Hounds from Saintes, 466 miles, both in 1981. Unraced in 1980, Dail's Delight won in 1979 as a young bird 1st Club, 1st Worcester Fed from Weymouth, and all the time Robert has been going from success to success with his fabulous team, small in number as they are, he has yet to find a cock to emulate the performances of Dail's Delight.

Similarly the late John Langstone who was a confirmed Widowhood racer found greater fame through the racing exploits of a nestmate pair of hens namely Sister, 5th Open Bordeaux Grand Nat and Cinderella, 1st Open and King's Cup Pau, and between them a number of other exceptional performances, although Sister's racing career was brought to an early end owing to her habit of fielding and subsequently being poisoned from which she succumbed. Looking back through the long and glorious history of the British racing pigeon a considerable amount of that glory can be attributed to the feminine sex. It certainly makes you think, particularly after you have looked and digested the latest publications of the Fancy Press, it is no small wonder that potential newcomers to the sport are smaller in numbers than they once were because I seriously believe thay become confused with the so called technicalities of pigeon racing, pigeon housing, and all the so called gimmicks and gadgetry that we now see being advertised, and considered imperitive to success. Together with the very high costs involved and the prices asked for stock today they would make the boldest cringe at the thought of progress, like the disappearance of that delightful house in Cherry Orchard, Old Hill, Staffs!

The problem now is that the younger fanciers and newcomers, few in number as they are, are being preconditioned into the belief that only cocks should be raced, and that only certain strains should be kept. They are now being conditioned commercially into the belief that many and various electronic gadgets are imperative if they wish to become successful at pigeon racing. All this is very bewildering, confusing and expensive. A good clean well-built loft is an expensive item, and provided it is cleverly yet simply ventilated will provide you with a first-rate environment for your racers, in the same way that Bob Ashman has applied to his very simple yet healthy and straightforward loft layout that has helped him to win many races. A dry loft is important however humble it may be. A loft without draughts and always dry is a well-ventilated loft and one that is without any evidence of objectionable odours. Many of the great and glorious victories of the past were achieved with soundly bred pigeons kept under such simple conditions of environment. The essentials of patience, observation, and dedicated determination are the hallmark of a good fancier. To the young fancier my advice is do not be frightened into the belief that you cannot become a successful fancier unless you can pay hundreds of pounds for your initial breeding stock and equally have to buy a very expensive loft to which you must add all the confounded gadgets and electronic amenities now before the Fancy.

JANUARY

Many of the greatest performances ever achieved in the British Isles have been achieved with pigeons that have been bred for distance through a succession of trial and error. Ever testing one's sense of stockmanship through the basket.

It is a mistaken belief that money can buy success. To be a successful racing pigeon fancier you have to learn to appreciate the skills required, and such requirements cannot be purchased, they have to be learnt through observational experience. The successful breeder is a person who is fully prepared to work hard in the production of first class stock and this skill cannot be obtained overnight. It takes time, in fact, many hours of dedication and patient observation. If you desire to obtain such skills they will surely be seen to exist the moment you produce a consistency of type and performance over and above your previous efforts. Whatever one may think, or be persuaded to think through cleverly worded advertisements, you must first of all learn the business of keeping, breeding, and the training of pigeons in the simplest possible way. Learn to think for yourself rather than allow yourself to be manipulated by those whose sole desire is to make as much money as they possibly can.

The sport must go on, and it will not continue to prosper unless the novices, and newcomers, and the inexperienced are encouraged to learn about the skills required be become a successful fancier well before they part with their hard earned cash. The loft is of course all important. Without a well designed loft you will never be a successful racing pigeon fancier. Never damp, never cold, never too hot, always well ventilated. It has been noted and written that many bakers during the long and brilliant history of the British racing pigeon have been successful fanciers and this is undoubtedly true. Indeed it was true as long ago as 1883 to 1886 the respective years my father and mother were born that a famous baker by the name of J W Barker, then living at Reading, enjoyed a tremendous run of brilliant successes winning around forty 1st and 2nd prizes, 26 of these being 1st prizes, and of these 11 were 1st prizes from French race points. Interestingly from early volumes of the Homing News one of those actual 2nd prizes was won in a race from La Rochelle in 1886 in the very race that was won by the famous J W Logan with a pigeon that for ever afterwards was known as Old 86 which has ever since adorned the front page of The Racing Pigeon.

Although many believed J W Barker was Berkshire born and bred, this was not true for he was from Essex, in fact, he was born in that county. J W Barker also traded as a baker in Bishop's Stortford. This very successful baker and fancier also traded at Staines and brought out the biscuits known to children as "wafers" around the year 1912, but were originally introduced by Mr Barker as "orange and lemon" dessert biscuits. Later still owing to their success the rights of manufacture were later bought by the famous Huntley & Palmer firm of biscuit makers, they also bought Mr Barker as well, for he went to work for Huntley & Palmer and remained with them until his death in 1887. This was where he established a highly successful loft of pigeons, and no doubt the reason he was known as Barker of Reading. His most famous pigeon was a bird named "The Gooose" and I well recall my father once owning a grand-daughter of that particular hen that he purchased for five shillings! Actually it was one of my father's 21st birthday presents, and five shillings was quite a sum in those days. She was about eight years old too.

The J W Barker family was based upon Belgian stock that he obtained through the agency of the famous Northrop Barker who was well known as the sole agent for J W Logan in all matters relating to the purchasing of his foundation Belgian stock. The best of these imports for J W Barker were a Grooter cock and a hen bred by the celebrated Servais, the strain that helped in the highly successful establishment of the Stanhopes. As a result of the mating of a Grooter cock to a Servais hen the famous hen The Goose was bred. Although several other very fine pigeons helped to further the strain of J W Barker, such as The Ghost a pure white cock and another wonderful hen known as Shotover. The Goose was the most successful pigeon in his family according to the final sale catalogue. She was a blue chequer hen bred in 1880 and the majority of the wins obtained by her progeny made up the major part of a somewhat sparse catalogue, and she made the third highest price, nine guineas in 1887, when they were sold by Thomas May the leading pigeon auctioneer at the time and sold at the Public Hall, Upper Norwood on 16 November 1887. The 191 birds listed realised a total of £335. Top price was 17 guineas and lowest price three shillings for a self-black hen without a pedigree.

If you were to multiply these figures on the basis of around 28 times the total sum realised in 1887 to make a comparison with present values, it will give readers a fair idea of the value the Fancy placed upon the head of J W Barker of Reading family of racing pigeons. A pretty Penny indeed. Furthermore the Fancy of that time were right to do so, as events proved in the after years, for many of the most successful pigeons produced between 1888 and 1914 were descendents of the pigeons sold at that memorable sale in Upper Norwood. However, it is not for me to mislead since it is also history that J W Barker also used very successfully the J W Logan pigeons. Indeed two of the highest priced birds that made the two highest prices above that obtained for The Goose were two birds bred by J W Logan that the famous Reading fancier had only purchased the previous year in 1886 at the J W Logan sale.

One of these was a daughter of Logan's Old 86, top priced pigeon at 17 guineas at Upper Norwood. The other was a B Ch P C, Logan's 322, bred in 1884, winner of 1st prize Cherbourg, and a son of one of the most influential of the Logans, B16, which was a sister to N Barker's famous Montauban, sire of Old 86. This pigeon was second highest priced pigeon at the the J W Barker sale. There were six or seven other J W Logan bred pigeons listed, so it is possible this too helped in the successful disbursement of the genes collected and colonised with the J W Barker family that enabled them to produce so many good pigeons for others in subsequent years. Doubtless the early demise of the Essex-born baker, who emigrated to the West of England prevented his family of pigeons becoming even more successfully established.

Before finishing I will summarise, at least it will assist the novices and lesser experienced without subjecting them to the tantalising possibility of forcing them to "read between the lines". The essentials to the greater success of managing a loft of racing pigeons resolve themselves as follows. In fact, these are the golden rules as they have been since the early years of pigeon racing. The greatest handicap is to have a loft that is badly ventilated. It is equally as detrimental to overcrowd a pigeon loft as all too many are won't to do. In fact, overcrowding is for many more or less a habit, and a bad habit at that. Making haste too quickly is also a sin. Trying to win from the very word go without any consideration for the youthfulness of the babies is a deadly sin, especially for the newcomer who really should try to restrain his eagerness by tempering with a considerable amount of the greatest of all human assets, patience. Badly ventilated lofts are the most certain way to encourage disease — even epidemics. Pigeons of all breeds and particularly racing pigeons can become most depressed and unhapppy in an overcrowded environment. The trust that many place upon antibiotics in order to balance such shortcomings is dangerous. It has been written by some that overcrowding is the major cause of flyaways with young birds. It is certainly a logical possibility. One of the major causes of lack of success with young birds especially is you get the idea that young birds must be kept short of food in order to control their trapping. Much of the real trouble is that young stock are not fed well enough at the crucial time of their development. It is equally wrong to underfeed as it is to overfeed. Remember as I have written so often over the years it is love of home that impels our beloved racers to race home. No home is a happy home if it is overcrowded and the occupants are underfed.

Paramyxo
There is not the slightest doubt that paramyxovirus is well and truly established despite the efforts of many fanciers who have steadfastly

MONTH BY MONTH — in the loft

followed the advice given by the Ministry of Agriculture, Fisheries and Food (MAFF) to vaccinate. Fanciers are repeatedly calling upon me to supply vaccine although by law I am unable to, since I am not a qualified veterinarian, or pharmacist. However, owing to many enquiries made of me I know only too well of the numbers who have this dreadful scourge within their lofts or believe their birds are infected.

Many fanciers who have faithfully followed the advice put forward by MAFF claim that for everyone who does carry out according to MAFF there are probably ten who do not. From the information I have, I would not be surprised if the number was even higher. On the other hand there are many who give the jab to their old birds, but who do not consider it necessary to jab their young birds! Further, there are fanciers who only jab once and never carry out the recommended second jab. Many others blame stray pigeons that return after a long absence. Further to this state I am aware of the enormous anxiety that is brought to bear when a stray pigeon is found by a non-fancier and the owner is bound by RPRA rules to have the pigeon sent back. All stray pigeons, whether they return of their own free will, or are reported, are potential carriers of paramyxovirus. There is not the slightest doubt that all pigeons whether they are lost racing, or return after an absence of time, be it days, weeks or months should be kept isolated. It will take time to prove whether or not it has caught the virus, and that will require several weeks of isolation, up to 35 days is quoted by those with academic and/or veterinarian qualifications.

Most fanciers who seek advice from me are given the following: Take daily note of the droppings of all birds kept, and particularly returnees lost from the loft, in training, through racing; or those you are obliged to send for through being reported through the Union, or reported through other channels be it the Police, RSPCA, or a non-fancier who has gone to great lengths to locate the owner of a stray pigeon; take immediate steps to isolate the returnee, and be fully prepared for a period of not less than 35 days isolation; and once you note loose watery droppings you can suspect your latest returnee to be suffering either from coccidiosis or internal problems, maybe enteritis.

In the case of the former you may have loose watery droppings or discoloured water droppings, or even loose greenish droppings, which are accompanied by a progressive loss of condition. In advanced cases you will note a loss of colour from the iris of the eye. All too often if it is coccidiosis, you will note a larger than normal amount of water in the crop, and little or no evidence of corn in the crop. Many years ago fanciers referred to this complaint as 'Going Light' and all kinds of treatment were meted out to the unfortunate victims. Today fortunately infected pigeons can be treated with Coxoid, and provided the complaint is diagnosed early most infected birds will recover after

JANUARY

a seven day treatment of Coxoid and the rate of one fluid ounce (28ml) to one gallon of water. It is more than possible to treat 30 pigeons with a single bottle. Make certain to treat strictly according to the instructions.

Having carried out treatment for coccidiosis, and discovered your bird, or birds, are still far from well, then you will have to consider other measures, even veterinary advice. Or you can judge for yourself by observing the following which I have learnt from continuous observation over a long period. Whilst it is not impossible for a pigeon to be a victim of coccidiosis, as well as enteritis (enteric inflammation) or maybe worms, either hairworms (Capillaria) or roundworms (Ascardia) in each one of these complaints there is a similarity of symptoms, but in each case a slight difference, that unfortunately I, like others, have had to learn the hard way.

Symptoms and their similarities
There is no doubt in my mind that coccidiosis will bring about a continuous loss of weight, accompanied by loose, and/or discoloured droppings, as well a loss of colour in the iris of the eye, and noticeably an excessive intake of water, at least this has been my own experience. But whereas the loss of weight with coccidiosis is comparatively slow, in the case of worms, particularly hairworm (Capillaria) loss of weight occurs much more rapidly, and as the weight is lost, so excessive thirst is noted. With worms old birds suffering from an infestation of worms can live for weeks, even months. And believe me many do, and all this suffering and discomfort during the racing season, thus preventing the victims from ever being able to manifest any indication of racing ability and with dire consequences to the owners' desire for racing success.

Vomiting is also a possibility when a pigeon is a victim of worms, particularly hairworms (Capillaria). Roundworms (Ascardia) do not as a rule cause the severity of symptoms as produced by hairworms. Again in dealing with the similarity of symptoms vomiting can and does occur with pigeons which have developed digestive disorders.

Whenever I observe a pigeon vomiting its corn, or any other substance it has consumed I immediately administer a sour crop or digestive pill. Most of those I have used are based on charcoal. Again as in the case of coccidiosis, worms, or canker (Trichomoniasis) all these diseases can produce loss of weight. Careful observations, much patience, and practical experience will enable you to treat your patients successfully, in the same way that a general practitioner can successfully prescribe for his patients as a result of his experience and knowledge gained from symptoms described by his patients.

The great difference of course, is that our beloved pigeons cannot describe how they feel. You, the owner/breeder, have to diagnose according to your own knowledge gained through experience, and this is not learned easily or speedily. Hence my desire through this chapter, as well as my love of pigeons to try in my own way to assist fanciers, as well as try to reduce pain and discomfort for our suffering pigeons.

There is not the slightest doubt that disease is a great drawback to success in pigeon racing. It is true that only fit pigeons win prizes, so we must do all in our own power to manage out charges. Health, good health that is, is of the greatest importance. The possibility of disease, and the seriousness of the consequences that follow a disease are of paramount importance in the mind of the experienced fancier Absolute success in the treatment of disease in a pigeon, or pigeons, is entirely dependent upon the correctness of the diagnosis and subsequent promptness of the treatment applied. Very few veterinarians are specialists in the diseases that occur among the ranks of the racing pigeon, at least not in the way the members of this august profession apply themselves in Belgium. Here and there you may be fortunate to come across a veterinary surgeon who is interested in the racing pigeon, if you do, you are indeed very fortunate.

Paramyxovirus and Polyneuritis symptoms
In advanced stages of both these diseases, there is undeniable evidence of similarity of symptoms. In both cases, there is a loss of the co-ordinating powers of the nervous system and the musculature. With paramyxovirus, as it advances its development within its host victims, within the same loft may develop a twisted head and/or neck; in severe cases, these will produce the most distressful appearances. This loss of co-ordination of the powers of the nerves and muscles will also be observed in polyneuritis or thiamine deficiency. Unlike paramyxovirus this can be helped curatively by a constant addition of vitamin B complex tablets for a period of at least 14 days, whereas to alleviate paramyxovirus you will be obliged to vaccinate as a preventative. However, if you have not vaccinated against paramyxovirus, and discover wet or loose droppings in your loft, then provided you are satisfied, as a result of a process of elimination (treated for coccidiosis, worms, or enteritis), the latter can be treated successfully with Furoxine (obtained only from accredited MAFF approved vets or pharmacists) then the chances of you having paramyxovirus among your pigeons is almost certain. Stray pigeons, can bring it into your loft, or late race returnees, too, hence my advice to isolate all such pigeons, including new purchases, or presentation birds, for an observation period of not less than five weeks (35 days) period. The Ministry of Agriculture, Fisheries and Food (MAFF) is in favour of a dead vaccine (inactivated vaccine) as opposed to a live vaccine. The

Frank Hall's first loft with an Evangilisti trap, this was built in 1921 at the rear of 163 Blackhorse Road, Walthamstow. Frank's father was a mutton and lamb salesman at London's Smithfield Market and the loft was built from pork lion and kidney boxes.

Ministry maintain they have very good reasons for the effectiveness of the Glaxo manufactured Paramyx-1 vaccine, and in its very efficiently prepared pamphlet entitled Pigeons and Paramyx-1 (the registered trade name for their vaccine) provides comprehensive answers to questions asked by racing pigeon fanciers.

If you consider you have contracted paramyxovirus within your colony, then you should forget about vaccination, at least for a period of ten to 11 months, otherwise you will intensify the disease by injecting another dose into the system. Once you discover you have contracted paramyxovirus, you will be best advised to keep your pigeons confined. If you also diagnose paramyxovirus in a particular loft (for those who have several as so many do these days) or a particular section, then it will be wise to confine the inmates together whether they include both sexes or not, for at least a period of not less that 35 days but preferably 70 days. Wet droppings will disappear completely after the 70-day period. All droppings must be thoroughly destroyed through burning.

An unorthodox treatment
There is little that can be done once the virus has been contracted other than treat your pigeons with all the patient nursing commonsense dictates. Or you can destroy all those that suffer most, as so many have done rather than vaccinate. I recalled a long time 'phone-in' friend Bob Burrows of East Acton, who still flies with Acton NR club as Burrows & son. Both Bob and his father, William, were among the founder members of the club, and Bob has continued to maintain the original partnership's membership to this day, although his father has been dead since 1947! Bob phones me from time to time, and I recall Bob ringing me after I had published some years back my own nightmare experience with paramyxovirus. Bob is always ready to help and advise if he thought he could help a fellow fancier, so I thought it is about time I phoned Bob because, although anti-vaccinator that he is, he has a theory, with well kept notes to back up his opinions, concerning the use of an oil that his father used in his business.

The ever helpful Bob readily offered the following observations: "Whenever the unmistakable signs — twisted neck, shaking head, twitching movement of head, or unable to focus their sight to pick up corn, or wet droppings, then immediately administer ten milligrams of ground nut oil down the throat of infected birds (2 x 5ml teaspoons)." Probably it mostly is a two-handed task in order to carry it out effectively and with the least discomfort and anxiety to the patient. It means also obtaining a safely designed eye dropper, one with a rounded nozzle, as opposed to one with sharp edges, like some that I have noted. "Ground nut oil is quite wonderful" states Bob, "this oil will also make up for vitamin B deficiency." Continuing friend Burrows says, "It helps to give relief if you gently help to disperse the oil with a gentle massage of the neck, and also back of the head," adding, "it is important to administer with great care to avoid the feathering becoming too messy." Bob is adamant that ground nut oil helps to alleviate the twisted neck.

For those who have not witnessed severely twisted necks then let me assure you that the first sighting is a gruesome spectacle. It was for me a most awful sight. My own initial observations was the upward, sideways turn of the head. The kind of movement you have doubtless noted when your pigeons are out on top of the loft, especially young birds who have for the first time seen a hawk or other predatory birds, above, high in the sky. My 'phone-in' friend Bob further adds: "Whatever you do regularly, or even irregularly by way of soluble vitamins, or vitamin pills, stop using these immediately, once you introduce ground nut oil."

In one of many phone calls that I have made since concerning the use of ground nut oil Bob states that he himself experienced paramyxovirus in 1984 and 1986, and throughout these two periods maintained a day-to-day note of treatment and progress. Aware as he is of the Ministry of Agriculture, Fisheries and Food (MAFF) directive regarding vaccination as the only effective way to deal with paramyxovirus, like many fanciers even now he is totally opposed to vaccination preferring to deal with the virus his way.

Previously in the building trade, Bob became involved with the family business of fried fish & chips soon after his father passed away, and still is, hence his awareness of ground nut oil. Bob's father was always trying out various ways and means to improve racing fitness, and the feeding of the race team as well as the effectiveness of feeding for quality stock bred from the very few stock birds, mainly the strain of Soffles, that his father established his loft of racing pigeons on. Bill Burrows discovered the usefulness of coating his corn (feed) with a very thin film of ground nut oil, and used it whenever he thought it necessary to boost his race team long, long before paramyxovirus was known about. Because of his father's discovery of the usefulness in various ways of ground nut oil it prompted Bob to experiment with ground nut oil in the treatment of paramyxovirus. As a result of his experimentation Bob is now convinced more than ever that the administering of ground nut oil really does help to alleviate the dreadful nervous reactions that paramyxovirus produces.

Bob Burrows meticulously records his observations in written notes on all that he observes in his pigeons, in spite of his utter dislike of writing articles or even letters! Regarding his latter dislike I have been greatly honoured as you will note as you read on further. For business reasons, training and racing is somewhat restricted. The family of pigeons were based on the early Soffles, a popular and highly successful Belgium strain founded by Mons Soffle of Antwerp many years ago. To the Soffles were added the early Stassarts and lastly some Shearing Logans. The Soffles in particular were most appealing to William 'Bill' Burrows, and they flew with success despite the initial promotion of a fried fish & chip shop business, in East Acton Lane. The building was actually a pub, The Bulls Head, from whence the very first meals of fish & chips were served by William Burrows, and upon which the present business now stands, replacing the old Bulls Head.

Ground nut oil
Bob quoted an incident of some years back that set him thinking. A number of his pigeons were vomiting badly, so he phoned a veterinary executive who advised Bob to put them down. Bob's reply and thoughts are best left to one's own imagination! After deliberating Bob decided to administer his father's old stand-by, ground nut oil. The result was almost miraculous. After a few hours those that were more seriously distressed had improved. So Bob treated all his birds, and within two days the entire team were transformed. His father's old remedy had saved the descendants of his favourite strain of Antwerp Soffles. On another occasion Bob had two pigeons return from a race that were very poorly, and in Bob's opinion were suffering from poison. Within two hours of receiving a dose of ground nut oil the pigeons were transformed. Ground nut oil is not difficult to obtain. You will find it on the shelves of the main supermarket self-service stores, not very expensive either at around £1.35 per litre bottle. Bob writes, "more and more fish friers are going over to ground nut oil".

Further to the many telephone calls that have taken place between Bob Burrows and I it is my rare privilege to be able to include the major part of his one and only letter to me concerning his experience and experimentation with ground nut oil and paramyxovirus. Bob writes as follows "Reference the use of ground nut oil (peanut oil) to prevent the spread of paramyxovirus in the loft after the birds have been attacked. This is not a cure of the disease, simply a means of preventing it spreading to those birds that have not contracted the virus. There is a time lapse of weeks between infection and its full development. Symptoms to look for include squrting droppings, lack of muscle tone (body seems spongy), head twisting, loss of optical focusion and consequent inability to pick up quickly, and any erratic behaviour.

"I am convinced the disease is transmitted orally and contracted in the following ways: In race baskets, or crates, when mixed with birds that have contracted the disease during racing, or group training, especially after being fed in the crates during holdovers, or confined in crates or panniers in distance racing. Furthermore when the birds are flying from race points, or even flying around home and mixing with other fanciers' birds, they can contract or pass on the disease through the excretions from birds who defecate squirty droppings when in flight and their droppings will quite understandably fall and hit those in the rear.

"Obviously the feathers of the pigeons immediately behind easily become soiled with the loose excretions. Naturally these particular birds when preening and cleaning their feathers, the transmission of the virus is feasible — the contaminated dried excretion is therefore easily transmitted to the crop of the non-effected bird. Often too returnees from a race, especially young birds that have not been vaccinated, not also to overlook stray pigeons yet another source for the introduction of the virus. Visiting fancier's footwear is suspect too especially those with a deep tread. Sparrows and starlings getting into the loft for water and nesting material especially feathers are possible 'carriers' of the virus through visitations to contaminated lofts". Bob Burrows also remarks upon the fewness of sparrows these days!

Continuing Bob outlined his observations and the application of ground nut oil (which are included as written). these are itemised.

"1. Take away all water for at least four hours. The reason the birds go for the water is to obtain relief from the inflammation caused by the germ attacking the digestive tracts, the water has a cooling effect. (*This is also a symptom brought about through coccidiosis*—FWSH.)

2. Give all birds four (4) full eyedroppers (10ml) of ground nut oil, by easing the bird's head forward and up inserting the full eyedropping into the bird's crop approximately 2ins, squeeze the rubber bulb and keep it compressed as it is withdrawn. By wiping the eyedropper after each filling before re-inserting, very little, if any oil will be seen outside the beak.

3. Gently massage the bird's crop, try and get the oil to work around the bird's crop and upwards towards the back of the head. The object is to form an oil barrier between the crop and the spinal column to stop the germ attacking the nerves therein, also to sooth the inflamed digestive tracts. Repeat the above every four days until you feel you have the attack under control, the only difference being that you then use only two eyedroppers (5ml total).

4. Disinfect droppings boards, perches, and floor daily. I use Vykil as it smells better.

5. Do not use vitamins in the water as these contain substances that may reduce the oil barrier thus defeating the real object — the protection of the nervous system.

Finally always keep a small bottle of ground nut oil and an

eyedropper handy, the oil will keep for about a year. I have used it for many years for a variety of reasons. Pigeons that have returned home suffering from poisoning, or enteritis and digestive troubles. Within a matter of hours the victims are back to normal — it has always amazed me every time I use it".

Since the above letter was received, and further phone calls to Bob Burrows I have learned that Bob has used ground nut oil to quell an attack of sour crop with great success. When you consider all that Bob has outlined in his very well meaning efforts to help both pigeons and fanciers it's not a bad record for ground nut oil to say the least.

Bob Burrows has gone to a great deal of trouble in a sincere effort to help those fanciers who are suddenly confronted for the first time with paramyxovirus, to give it its complete veterinarian name. Bearing in mind there is still a great deal more we do not as yet know about this terrible disease. Today we know the virus exists, and kills including champions and that we are advised to vaccinate. There is a great deal we do not know since to a large extent there is a constant air of secrecy. The problems it creates, the experiences gained by fanciers should be shared. Let us hope and trust that in due time the Fancy will accept a stringent vaccination policy. The sooner this is made law the better it will be both for the pigeons as well as the hobby of pigeon keeping in all its aspects of racing, showing and fancy pigeons.

Regarding ground nut oil according to Bob he has had great success in several ways as outlined above. I was particularly interested, even amazed, in its use to combat sour crop. Not to overlook its possibilities to overcome vitamin B deficiency. But do bear in mind that what has been advocated by Bob Burrows has been recorded in an effort to help those who suddenly discover to their alarm that one of their pigeons has contracted paramyxovirus. Remeber it is no use at all to vaccinate a pigeon with obvious symptoms of the virus. But there is nothing at all to stop you having the rest of your team vaccinated.

Stick to any of the vaccines approved by the Ministry of Agriculture, Fisheries and Food which are all dead vaccines. How much simpler it would be if United Kingdom and Irish fanciers were allowed to use the Continental vaccine so freely sold over the counter at pharmacists on the Continent without fear of hindrance. But the MAFF is sternly opposed to the use of a live vaccine. After vaccination administer vitamins for several days, at least five consecutive days.

Remember ground nut oil contains vitamin B and if used in conjunction with bear yeast powder which is rich in both amino-acids as well as B vitamins will help your vaccinated pigeons enormously. Simply add ground nut oil to the feed mixture (corn) at the rate of a level dessert spoonful to 14 pounds of feed mix (best to use rubber gloves) especially if your mixture contains barley, and then dry the mixture off with bear yeast powder. Make your complete feed up overnight. Next day the mixture will be just right for use. Pigeons will soon get used to it. But I do not advise you overfeeding, or hopper feeding and leaving it in large quantities to become scattered and consequently soiled, that's not the way to go about it. Be sensible about the way you feed your pigeons, cleanliness is next to godliness. The basic requirements of the average pigeon is one ounce per day, until your pigeons are working hard with training and racing when they will require more even up to one and a half ounces a day. The art of feeding is the secret of success.

Racing pigeons are simple enough creatures
According to my own studies over a long period of time pigeons are motivated entirely in their activities according to their desires or requirements of the moment. They are like children, creatures without artfulness as young birds, even as yearlings. To be more specific they are without treachery, and certainly without cunning, or even deceit.

Like children too, they are influenced tremendously by their environment. It is also important to be reminded that we create the environment for them. It is equally important to remember that the environment can change and the individual accordingly.

A good environment implies that the air our pigeon's breath is clean and as much as possible without dust, and most certainly not overheated. The water must always remain clean, especially not polluted with excretions, or any undesirable bodies. The food they are given must also be sound and clean. The education and guidance they receive from the very beginning must remain regular and disciplined. This will ensure they come to order. Put another way young pigeons must be trained to obedience, otherwise you will not succeed when you require immediate obedience from the leading birds when they arrive home from a race. Remember we are racing pigeon fanciers and, therefore, we are dealing with potential racing pigeons that have to be educated to race home in the quickest possible time. Furthermore when they reach home they must be ready to trap without hesitation otherwise precious time will be lost.

The environment includes the most elementary basics that we are inclined to overlook. Possibly for many who fail to recognise the importance of a warm nest, or a comfortably clean nest box. Even the quality of the eggs is influenced by the enviroment. The comfort too of the parents during the period of incubation is of great importance. The preparation of the parents prior to being mated, and their subsequent treatment, and management are of great importance.

Quality feed, clean grit, clean minerals, fresh vitamins, and recognition of the basic rules of hygiene will play an enormous role in the production of quality eggs. All of the above observations are part and parcel of the environment. The surroundings in which we enshrine our pigeons will rightly or wrongly influence tremendously the future character of the individual from the moment of fertilisation of the egg. The environment is so very important it can, if studiously considered, improve the good pigeon, and steady the wild. Although the environment cannot alter the basic instinct of the species it can undoubtedly influence the nervous disposition of the individual. So much so that the powerful influence of the environment (surroundings) can reduce the tameness of the tame, or increase the wildness of the wild that a poorly managed loft, despite having been founded upon good pigeons, will remain unsuccessful. Or put another way an unsuccessful loft of racing pigeons, is one in which the occupants suffer from lack of simple commonsense, elementary stockmanship, and the basic principles of what constitutes good management. Therefore, until the environmental basic principles of hygiene, loft comfort, feeding, watering, early discipline, training and creature comforts are improved so will the loft remain constantly unsuccessful. The influence of the environment is so important that it can improve beyond all conception the innately superior pigeons. Yet the basic principles of a sound environment are in themselves quite simply a matter of commonsense. The loft itself is important, but does not have to be palatial.

Further to the subject of environment, the matter of tameness is often overlooked by the novice. Whenever novices approach me to seek advice on the obtainment of foundation stock I invariably advise them to obtain adult stock from which they can breed their very own squeakers. The policy is twofold. It gives the newcomer a great opportunity to learn the need to know about the rearing of squeakers from the beginning, as well as getting to know each pair of breeders. This is the finest way to obtain first hand knowledge of the basic rules of the environment.

Once you have obtained your initial stock, it will enable you to learn to handle your pigeons. If you have gone to a reliable fancier, they will no doubt have given you advice on matings. This will enable you to mate your pairs accordingly and once happily mated they will soon learn to settle down in their nest boxes.

FEBRUARY

Loft Conversions — Early Breeding — Colour Breeding — Veterans — Fred Marriott Old Billy and Forlorn Hope

To help you to produce tameness in your squeakers make it a rule never to place nest boxes too high, or in a position that prevents your squeakers from getting to know you and your voice, and your movements about the loft. Make it a habit to talk to your pigeons. Stroke the babies once they are fully fledged. Avoid sudden movements. Remember creature comforts, a warm nest, a clean nest box. This comes within the scope of environment, and therefore is fundamentally of the greatest importance.

A novice fancier seeks to learn how to become a successful fancier and reach the top ten in the country in the shortest possible time? What a question! However, it conveniently fits in with my reference to the question of environment. The answer to my novice friend is "a great deal of hard work", equally lots of patience, a well-designed loft, and well bred pigeons.

His immediate reply was "And how much will it all cost?" Yet another most difficult question to answer. Finally he asked, "And how long will it take"? Regarding the matter of time. It can take several years, although probably shortened if you develop commonsense like patience. This is more or less a contradiction but experience takes time. It will also require the utmost powers of observation. For those who have an inborn sense of stockmanship they are the most likely to succeed quickly. Or should I state they will the more likely be counted among the newcomers who become successful in a shorter period of time compared with others. Having written thus, it would appear to me that anyone who shows a desire or an interest in the keeping of livestock possesses the earliest form of stockmanship.

Unfortunately among the very young some there are who seeking to keep pigeons are at first adamantly denied parental permission to keep pigeons! For many and various reasons it seems. However, for many parents it appears the main opposition is based upon a silly superstition that "pigeons are unlucky"! Seemingly the other main reason is "What will the neighbours say!". Assuming parental opposition is overcome, that is when the design of the loft is so important. It does not have to be elaborate. But it should not appear as an eyesore. Simple and clean-cut is best and painted white, or a nice green; oak-stain too is pleasing. It matters not whether it be a wooden structure or brick built.

Some, more fortunate than others, may have the opportunity to convert an existing brick building. This would prove a boon should such an opportunity present itself. However, for the majority it would more probably be a wooden structure. Here again existing buildings could be utilised. You can, if you take the trouble, search the advertising columns describing portable buildings designed purely for poultry. It is surprising how easy it is to convert the front of such a building for use as a pigeon loft. They are a great deal cheaper. Of course there are also a great number of very fine sectional pigeon lofts now manufactured as The Racing Pigeon journals both weekly as well as monthly illustrate. Nonetheless you can with patience and commonsense quite easily build a loft with various kinds of useful materials that will help you to get started.

With a modicum of commonsense it is quite surprising what one can do by conversion. The point I wish to emphasise is that you really do not have to build a palatial building. The basic requirement in a pigeon house, is that it should afford absolute shelter in wet weather. Also the building must have sufficient ventilation for the inmates when they are confined to the inside of the loft. Also according to the numbers required enough room to avoid overcrowding. To house too many is both costly and most unwise.

Many years ago I converted the hay loft over the stable where I kept my first riding horse and it proved an admirable pigeon house. The roof was sound, and prevented dampness after rain and/or snow after it had thawed. It was dry in the winter and cool in the very hot weather. The floor was sound and the stable smells did not appear to upset the pigeons in the slightest. However, not everyone is so fortunate to have such facilities. Although I do not wish to imply that my facilities were sumptuous. Not a bit. The stables I refer to were situated in Manor Park, a few yards away from the Three Rabbits Hotel, in Romford Road, No 829, where I lived and worked as a butcher

FEBRUARY

and that was in 1931. In 1934 I bought a butchers shop in Station Road, Forest Gate, where I converted an old galvanised roof pickling shed (pickled pork and salted beef) into a pigeon loft. All I had to do was design a front and one side.

The interior fittings were mainly Tate & Lyle sugar boxes that cost the princely sum of a shilling (5p) each. These really were ideal for nest boxes. You could buy damaged ones for much less and these provided the material to make the box perches and food hoppers. In fact, it was the size of the famous Tate & Lyle sugar boxes that dictated the measurements of the average Natural-type nest box front that remains fixed in the minds of the manufactures of wooden as well as plastic nest fronts designed for the Natural type racing fanciers and those who breed on a commercial basis to this day. A perusal of the advertisements of those who produce pigeon loft fittings will readily prove this.

In the early twenties when I lived in Walthamstow, Blenheim Road, to be precise No 46, I converted a large dog kennel into a small pigeon loft. By the time I had improvised it really looked first class. In 1926 when I lived at 230 Evelyn Street, where I also worked as a butcher, I built a really first class pigeon loft out of Argentine pork loin boxes. It was superb timber, and as each box was carefully taken to pieces, even the nails were straightened where necessary and reused. It was my pride and joy. The late Les Gilbert of Forest Hill, when paying a visit took a photograph, unfortunately I have lost it. However, the point of my reminiscence is to enable my young novice friends to be prepared to seek out a building that will stand conversion, or find material that with a little imagination can be used in the building of your own pigeon house. Maybe it would prove of great usefulness to novice fanciers, or those just entering the sport, if a photographic competition of DIY lofts could be promoted. Each entrant to be persuaded to give a list of materials used. It would I am quite certain prove most illuminating for those whose finances are restricted.

As I continue with this chapter I am mindful of February 14, Valentine's Day, the date recognised by most ornithologists and Nature lovers as the day when the bird world in its natural surroundings turns its thoughts to procreation. It is, therefore, assumed by hundreds of pigeon fanciers to be the ideal time to mate their pigeons. It probably is but with reservations according to dates applicable to the two main race routes. From past experience I consider it an ideal time to mate North Road pigeons, and also stock birds. For the South Road races I have learned to appreciate that the middle to the end of March is more suitable as it also is for 500-mile NR races in July. However, the view from out of my office window may be beautiful, but with a blanket of pure white snow all around it is most chilling. Personally I prefer to wait until the weather is warmer — much warmer in fact.

MONTH BY MONTH — in the loft

The very thought of spending hours and hours in the loft with temperatures below freezing, and the prevailing easterly winds leaves me cold. It even makes me feel colder thinking about it. Nonetheless I do know many who have squeakers in the nest, and some even ready to leave the nests. A great deal depends upon one's plans and ambitions. To be perfectly honest you can breed good sound youngsters at any time of the year if you know your job.

During the 1939-45 war, we of the National Pigeon Service, were even commanded to produce squeakers for war service all year round! And we did, with a great deal of pride. And as the records show, our efforts and dedication and pride in our war effort, as a voluntary body, helped to produce many fine squeakers despite the long dark winter blackouts, especially during the very severe winter of 1941. It makes me shudder even to think about that particular winter.

My pigeons bred early that year and required my attention to their needs. In particular their needs for a constant supply of fresh drinking water during the big freeze to enable the parents to carry out the rearing of their squabs without let or hindrance. Actually these squeakers were the result of love matings. The squabs were hatched out on the floor on deep powdered, perfectly dried-out droppings with plenty of short-cut straw and even an old hat that I uncovered by chance. From time to time, about twice a month I sprinkled some floorwhite to which I thoroughly mix a very small dusting of bacteriacidal blue powder. The floor is always dry, my loft is ideal in this respect. Truth to tell of the three pairs of eggs laid, five hatched, one egg was broken, and the five squeakers reared are without any doubt five grand babies.

All the winter breeding activity much more recently was largely prompted by Jackie Passaway who mentioned that he was going to present the prizes at a big Federation prize giving down Essex way and mentioned to me the exploits of a very young fancier by the name of Jason Roberts of Takely, a very proud 14 year old. I was so impressed that I immediately told Jack that I would breed for young Jason a pair of squeakers. It was actually planned by me way back in early November when I promised myself I would breed a few squeakers for members of Smithfield club. Planning is probably the wrong word but know myself well enough to realise that I will not miss an opportunity to save what I consider are worthwhile eggs if at all possible. The stock produced can be placed to advantage as in the cases cited above, even if it adds to my workload.

My personal and work pressures are the main reason that a few love matings have taken place. This method is not my normal procedure, in fact, far from it. When after careful examination of identity I discovered couples that were considered ideal, as well as being of the same family or strain and the squabs eventually produced were really

FEBRUARY

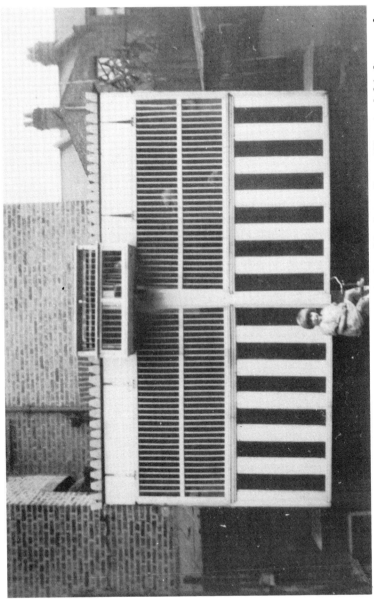

The loft at 46 Blenheim Road, Walthamstow in 1924 with Frank's sister Joan. The stock birds were housed behind the loft in an enlarged converted Airedale terrier kennel.

first class. I took the opportunity afforded to keep those self-made promises to provide presentations for a few friends of Smithfield Club.

One of the pairs reared so very successfully was a pair of Busschaerts, they are two splendid black chequers. These I earmarked for my old friend and Smithfield Club member Henry Adams of Tottenham. The second pair were from a 1985 bred pair of Vanhee cum Roger Vereecke, both parents being down from outstanding Stichelbaut lines. This pair went to another fellow member of Smithfield Club in Mick Caccavone of Streatham. The other of the five reared went to Jackie Passaway, a friend of longstanding and of course a member of Smithfield Club. This squeaker is of Dordin bloodlines Sosie, Spahi and Vanhee. It pleases me to be able to relate that all five squeakers appear to be settling happily into their new environments, despite the fact that they were bred and reared in deep litter, or powdered dropping to be more exact. They also had to endure the most severe cold spell we have experienced at this time of the year for the past 23 years. If they continue to be cared for, well fed, kept warm and dry they will prosper as did those squeakers reared in wartime Britain from 1940 until 1945, that helped so successfully the combined Services and underground freedom fighters in France, Belgium and Holland in such efficient style as the War Office records have shown.

I am not because of the above opportunism, advocating breeding deep litter style, or even advancing the merits of early breeding, I am describing what has happened in my own loft as a result of circumstances. It would be foolish of me to condemn early breeding owing to the varied experiences I have known in the past, and that is this, you can breed pigeons at almost any time of the year, if it were as in the last world war considered so very necessary. You can rear pigeons in the most unimposing places so long as you care for them with sympathetic and compassionate understanding. Most importantly this implies the need for sound clean corn, and as I have found, a small amount of pellets will not come amiss. Small quantities of clean grit are important instead of large quantities that are left for days, even weeks, uncovered and contaminated as a consequence. Sprouting peas in the absence of green food are also an advantage. But especially important when the winter is at its worst is the constant need for fresh water that is drinkable rather than being frozen and thereby undrinkable for long hours at a stretch.

Nests too need to be free from smells and vermin. With warmth in the nest at all times you will rear the best possible youngsters. In short it implies the need for simple type attention with particular emphasis on water supplies. In plainer terms a good warm dry home, with caring parents that are helped in their hard-working endeavours to bring up their babies in an efficient and appropriate manner by

FEBRUARY

meeting their needs at all times. That really does sum up the successful rearing of healthy and indeed happy young pigeons. There is only one thing I would add to the above routine of caring and kindly fanciership which is always the hallmark of a good fancier, I am never without a tube of cod liver oil capsules, or halibut oil capsules in my loft coat, ready and willing to slip down each squeaker's gullet in a gentle manner when they are getting towards the weaning stage, especially to each squeaker as soon as they are removed from their parents. This policy I continue for several days after they are away from their parents, sometimes for a whole week. This has worked well for me over the years. Failing cod liver oil capsules you will also find that a mixture of whole groats, and pinhead oatmeal in equal parts laced with cod liver oil and allowed overnight to soak in, is a very good natural feed supplement for young growing stock, or you can use a well-known Hormoform supplement made by Harkers since 1954 or 1955.

Frequently at this time of the year I receive enquiries from fanciers both young and old who seek to know how best they can produce hens, rather than as many appear to do, produce an excess of cocks. Well I am very much afraid that I do not know the answer to that one. There are times when I really wished that I did; Overall I believe we tend to produce more males than females. Indeed I believe this is really Nature's way of helping towards the successful maintenance of the species. As a consequence males in the wild must fight among themselves for possession of the females, and only the strongest will survive to imprint their physical strength upon the breed. This could well be true of the racing pigeon. It is a view that I have held now for many years. Further to such thoughts I believe that when you frequently produce two hens in a nest, or two cocks in a nest, it is perhaps Nature's way of telling that a cross is needed. But I am getting away from the real question asked of how best to produce hens.

To be honest I really do not know the answer to that one, nor does anyone else for that matter. However, I do know that in certain pairings you could be certain that particular colour matings could tell you by feather colours what sex you have in the nest.

This might prove of interest and perhaps even help those who appear to produce more of one sex than the other or maybe retain the wrong sexes. Two red chequers mated together will be certain to produce mealy hens, blue chequer hens, even dark chequer hens, as well as blue hens, but of the last three colours listed you will not obtain cocks of any of these colours. I am very fond of a dark chequer hen from two reds, perhaps I have been lucky in this respect but it does not happen very often, at least it most certainly has not been my experience. If you mate a red chequer cock with a mealy hen you will obtain similar colours as with two red chequers mated together. If you mate a blue chequer cock with a red chequer hen you will be certain to produce red chequer cocks and mealy cocks, but only ever blue chequer hens and blue hens.

Maybe before proceeding further it is as well to reflect that two reds paired together will produce red chequer cocks and red chequer hens, although the latter colour and sex I have noted is less plentiful, also mealy cocks and mealy hens. And of course the same applies to a red chequer cock and a mealy hen pairing, but again only blue chequer hens and blue hens, including the occasional dark chequer hen will emerge from this colour mating as happens when two red chequers are paired together. From a mating of a red chequer cock to a blue hen you will get red chequers, blue chequers, mealies and blues of both sexes, and this applies to red chequer cocks mated with blue chequer hens. A cast-iron method of producing blue hens is the mating of a mealy cock to a mealy hen, but remember you can also obtain mealy hens, and only ever mealy cocks, never blue cocks. Sometimes you cannot be certain about the sexing of the mealies until they show some maturity, or when cocks are flecked with black markings at an early age, which does not always happen to nestlings. You must also be certain that they are black markings and not simply a darkish brown fleck that is often the character designate that signifies the said mealy is a hen. The greatest certainty from the mating of two mealys will be that the blue will always be a hen. Although very seldom have I ever mated two mealies together, but when I have carried it out as in the very popular matings of two blues together, I am always aware of the importance of good strong bars in at least one of the pair. In the case of the mealy's good strong deep (wide) coloured red bars and in the mating of two blues, good strong deep black bars, that is in cases of good bred pigeons that show a thinness of bars. Mostly I prefer to eliminate those that reproduce very thin-barred markings in both blues and mealies. The important points to remember when mating is that one is wise to mate a good strong colour to one that is perhaps lacking in depth of colour. For example, in the same way that I am also very keen to mate a dark with a light, viz a mealy cock with a good type dark chequer hen or dark velvet hen, or a black chequer hen; or vice versa.

With the system that is practised by most fanciers in most parts of the pigeon racing world, unless you segregate each pair of birds you breed from, that is to say that each and every pair you require to reproduce from is maintained in its very own compartment without providing any possible intrusion by another male at any time no one can be one hundred per cent certain of the parentage of any pigeon. The exception could be in the case of those fanciers who like myself value the offspring of first round youngsters above all others and only ever let one pair out at a time in each section until all eggs are laid.

FEBRUARY

This loft was erected close to the house at 230 Evelyn Street, Deptford, and was again built mainly from pork loin boxes. It was to this loft that Frank timed his first winner, a red hen known as No 19. Frank moved to Deptford in the summer of 1926, in time to breed for the following racing season and to help establish the Deptford Central FC.

Several of my valued pigeons that are getting on in years are confined to their own quarters. Two things are then certain, the parentage of the squeakers, and if the male is getting on, or even the hen is old or nervous at least there is no possible chance of either of the special pair being disturbed or bullied by interference by another bird when copulation takes place. All the above applies particularly to stock birds. Where one only maintains a totally flying out colony, and pigeons are flying fit, the danger of interference is less likely because no male pigeon that is worth his salt and flying fit will take his eye off his hen when driving. Even so from a scientific point of view the best and safest way to be absolutely certain of the parentage of any pigeon is through complete segregation of all pairs. But since few of us, in fact, very few indeed can afford such facilities we therefore have to rest content with what the majority have.

I have just remembered, a fancier of considerable success phoned me regarding a colour query. The question put was: Could a blue cock be bred from a pair of red chequers? The answer is, or at least the one I gave my caller was "not possible", as I have shown above. Similarly a fancier once asked me to help him solve a problem with the breeding of a red chequer cock that he had produced from two blue chequers. The answer I gave him was an emphatic "not possible". After several letters and phone calls, like the majority of fanciers his birds were not segregated, the bird in question was a second round pigeon, and there were three red chequer cocks housed in the same loft! Colour breeding is a fascinating subject but few it appears are keen enough to make a complete study of it, or even have the time, money and facilities to pursue the subject on a scientific basis. Finally on the question of colour and I mean this, from two blues you can only be expected to produce birds of the same colour, and of both sexes.

There can be no doubt at all that it would prove most useful to would be fanciers whose finances at first are somewhat cramped. Above all else to reiterate, whatever you do, DIY. Remember if you make certain your pigeon house is dry, and equally sensibly ventilated, you have acknowledged two basic requirements of a suitable loft. A sound roof, and a sound floor, are equally essential. This is the begining of a suitable environment, probably the most important word in the successful development of a family of racing pigeons. The loft is their home, and therefore, like your own home, must have all possible creature comforts. Make it pleasing for your neighbours to look out on to, as well as simple to maintain. Do not complicate the issue with too many unnecessary fittings, and complicated gadgets.

Veterans
An old timer who in 1990 called upon me is veteran Steve Tucker of Walthamstow, who could not get anywhere but for the goodness

of that splendid fellow, long distance racing enthusiast, Tom Watson. Despite then approaching his 87th year, Steve Tucker is now looking for a good cross for his old fashioned English-based family! How keen can you get! Well done Tom, for being such a good Samaritan, and equally congratulations to Steve for being so "doggone stubborn". Another veteran that springs to mind is George Swann, with seven out of eight from Pau NFC, who was then in his 81st year was still planning for both roads, including the Nationals! Another too that calls upon me now and again is veteran Bill Denyer who despite his 84 years still gamely fights on in a hobby that he took up very late in life, largely due to his son Peter's interest. And so it happens that Peter's interest was triggered off more or less through yet another member of the family, Bill's grandson, young Terry Yull.

All credit is due to these veteran enthusiasts. The pity is that so few younger fanciers are entering the sport despite the great efforts of those who are doing their utmost to help and encourage young newcomers to the ranks. Especially pleasing was Liz Hobbs MBE's efforts through Yorkshire Television. Maybe it will now be necessary to introduce a youth membership scheme whereby only half subs and a half nom fees and race fees too will be the order of the day. All this and well bred pedigreed young birds free, and breeders too, supplied by members who are prepared, like Jonathan Mee through Young Fanciers' Forum, to put themselves out in order to help the novices. It was pleasing to note "Presenting the Pigeon Sport" through the eyes of TV cameras by Robert Charles and John Helm was a great success. However, old 'uns will have to do a lot more than we are doing at present, if we are to encourage newcomers into the sport. Those outside the hobby of racing pigeons know little about this sport except by obtaining a copy of The Racing Pigeon or other pigeon papers, and since these only circulate within the Fancy you have a problem.

Pigeon racing is both fascinating and worthwhile but it is fast becoming a very expensive hobby. Like first-time house buyers, novice fanciers face high costs. Systems of racing have taken over, those with time and the money, coupled with the knowledge and the dedication to excel can do enormous damage in the demoralisation, racing-wise, of the novice. It takes a very determined youngster to accept a weekly drubbing from those whose sole purpose it seems is to dominate each and every race they enter as well as every club, no matter what the cost to the newcomer. Unless you impose limits of birdage, and this is not the best way to run a club financially since all clubs must pay their way, then limit the number of prizes a member can win, say two prizes only. Let all who are interested clock in as many as they want to, but at the same time reserve two of the eight diplomas for those who have not ever won at least three races. Every encouragement must be made to help the newcomer to stay in the ranks.

Specialist systems based on Widowhood in varying forms have taken over the sport. And because of this a complete transformation has taken place. It has created the need for many additional fittings, fixtures, and paraphernalia. For example, Widowhood nest boxes, Sputnik traps (which are really a more sophisticated version of the piano trap) and other contrivances. The systems have also increased considerably the price of pigeons. The quality of the pigeons appears to depend a great deal upon the number of first prizes a loft has secured in short to medium distance races. These lofts with a good position and overfly have the advantage, whereas in days gone by it was distance winners that mattered most. And in this the old fashioned Natural method provided novices with a chance to learn racing in a far less cut and thrust manner than prevails in clubs today. Novices need all the help possible.

Even in early February many will have begun breeding in earnest. However, specially for novices, it is early days and although the weather can be remarkably mild (at least it is down south) for this time of the year there is always still plenty of time to get some very hard weather in February. Nonetheless for the keen ones they will not easily be deterred, hence the following true story.

Fred Marriott

There are many who now and again have a go in putting together the odd pair of two, just for the fun of it all. Many years ago, it must be around 50 years at least, I recall the late Fred Marriott attended a London Social Circle gathering of distinguished fanciers in London, at the famous First Avenue Restaurant in High Holborn, and told the story of a very fine racer, that, like so many of Fred Marriott's pigeons, traced back to his famous Dreadnought. Mr Marriott in his recollections recalled a pair that he fancied breeding some early ones from, and duly allocated them the entire young bird section. A nest bowl was duly placed in the more sheltered corner of the loft, making certain the pair were really well looked after. Despite the severity of the weather at the time that pair of pigeons repeatedly removed the nesting material and took it into the trap where it really was cold and drafty. In those early days the traps mainly in use were the Evangelisti type, an extension trap. And there they laid a pair of eggs. Only one egg hatched but it turned out to be a very good young bird racer and finished up winning the last and longest distance young bird race. Later still that pigeon, 577, became the dam of Marriott's well known 1929 2nd Open San Sebastian-winning hen Nap, a very nicely made chequer hen. Nap lost the King George V Cup by the narrowest of margins. You never really can tell where a good pigeon is coming from.

FEBRUARY

If the weather is cold, or the situation of the nest is in a most exposed or unlikely position, even to the point of total discomfort, Nature will assert its authority and direct the natural parental instincts of procreation towards the creative comforts provided by close sitting parents! Sixty-five years ago pigeon racing was dominated by long distance racing under Natural conditions, and early breeding was not encouraged. Marriott's 577, the dam of Nap, was probably one of very few to breed a 2nd Open NFC winner from San Sebastian that was hatched on a cold day, in the earliest part of January 1925 after being hatched in an ordinary piano shaped trap with only the natural warmth of the parents to incubate her. Nap flew San Sebastian three times and took positions in the 1930 and 1931 San Sebastian races. St Mark, a light blue chequer cock ringed RP09Q3438 figured in the ancestry of Nap. St Mark, in fact, figured in the breeding of almost all of Fred Marriott's champions including the Champion Lerwick hen, twice winner of HM King George V Challenge Cup, from Lerwick, in 1920 and 1921, promoted by that famous club the NRCC. Both Fred Marriott's San Sebastian King George V Challenge Cup winners in 1924 and 1925 respectively, Triumph and Repetition, both descended from St Mark.

The noted blue pied cock Johnny White Tail, bred in 1917, a five-timer from Lerwick, four times in the prizes including 5th Open NRCC Lerwick, also included St Mark in his breeding ancestry. So too did the noted Paris Hen, a dark blue chequer white flight hen bred in 1927. She obtained her name by winning 1st prize Paris as a yearling in an extremely hard race. To emphasise still further how difficult the race from Paris to Stechford was, Mariott's winner was the only bird home in race time, she homed at 6.29pm on day of toss. This splendid racing hen also won the City of Birmingham £100 Paris race, through being nominated into this race, as well as winning all the pools, prizes, and all race prizes. Later still she won the Warwickshire Fed's big Open race from Pons, winning the race by two hours. Both Marriott's San Sebastian King's Cup winners Triumph and Repetition figured in her ancestry. Two other important pigeons in the F W Marriott family tree were Prince George, a pencil blue cock, and Princess Marina, a blue chequer and these also descended from the famed St Mark who in turn was sired by that remarkable producer, Dreadnought.

That is heredity for you, and emphasises the importance of such a discovery. Unfortunately few fanciers experience such good fortune. In 1915 Fred Marriott, due to the possible difficulties of feeding his pigeons during war time, included Dreadnought with several others that were killed and in his very own words "never made a greater mistake in his life". Marriott actually told me this many years later, at least 24 years after Dreadnought's death. Fortunately St Mark left his mark through his descendants. It is, however, a great pity that these are now so thin on the ground. However, there is bound to be someone, somewhere, who can pinpoint descendants. At least I hope so. Maybe we shall hear from someone.

Like so many highly successful fanciers of the past, Marriott's reputation was made by becoming highly successful in long distance racing with a team of not more than 12 pairs, including breeders. With producers like Dreadnought, followed by St Mark, you do not require large numbers. The famous Vic Robinson, another highly successful National ace and winner of King George V Challenge Cups, advocated a dozen pair of racers. Both the above fanciers had several things in common. They were not afraid to get up early. Both trained their first racers by means of a bicycle. Both won races in their very first year. Fred Marriott in 1899, Vic Robinson in 1911. Both excelled at long distance National racing and each only ever raced on the old fashioned Natural system of racing, both cocks as well as hens. Between them they accounted for the winning of no less than seven King George V Challenge Cups. Both fanciers were 16 years old when they won their very first race. Both fanciers also claimed to owe their success to line breeding to the best breeders by pairing the best racing cocks and hens (sons and daughters) back to the descendants of their finest foundation pigeon. In the case of F W Marriott to Dreadnought, the sire of winning pigeons with every hen he was ever mated to and in the case of Vic Robinson his finest foundation pigeon was The Dove, so named because of her small size. She was ringed 1502 by her breeder A H Osman, and was a direct daughter of Forlorn Hope and the Weinberg Hen. Almost all of Vic Robinson's best long distance winners contained the bloodlines of 1502. More of that in a later chapter but similary all Fred Marriott's great champions, including the winners of five King's Cups, traced back to his finest producer Dreadnought whose ancestry included A H Osman's famous foundation pigeon, Old Billy.

Old Billy and Forlorn Hope
It never ceases to amaze me how very interested fanciers are whenever I make the effort to recall champions of the past, especially whenever I research history with regard to the very few pigeons that really leave their mark on history. Among several interesting letters was one from Arthur Cassell of Sheerness, Kent. Mr Cassell writes: "As a schoolboy way back in 1951 I first kept pigeons for a year. At that time I saw a photograph of Forlorn Hope (1902-20) and have never forgotten it. I came back into the hobby last September buying a few late breds. From these I hope eventually to breed for the distance races. Your article mentioned that Vic Robinson's The Dove came from Forlorn Hope. I wondered if you would know if any fancier still has pigeons

descended from Forlorn Hope? If possible I would very much like to incorporate this historic bloodline into my loft of birds". Although Mr Cassell's enquiry is a real poser I do not doubt even in this day and age that if records have been maintained or can be researched through foundation pedigrees it will prove beyond doubt that descendants of Forlorn Hope still exist. Doubtless many will not believe this to be at all possible. However, the power of the genes are so strong they are practically impossible to destroy. Unfortunately many valuable genes are wasted.

Wherever you find a fancier who has a foundation based on the Vic Robinson strain, or that of the F W Marriott strain you will be almost certain to find descendants of Old Billy! Again this could also apply if you could trace the descendants of the one-time famous Harold Jarvis Osman-based family whose noted pigeon Marseilles King was an impressive descendant of the dark red chequer cock Forlorn Hope, ringed RP1902E1002 bred by A H Osman. His sire was a son of J W Tost's 142 and 182 a famous Tost pair whose children crossed well with the Old Billy genes. The Old Billy ancestry of Forlorn Hope was through his dam who was bred by the late Col A H Osman from Old Billy when mated to an N Barker hen. The number of winning descendants of Old Billy must have run into hundreds probably thousands including Vic Robinson's The Dove, F W Marriott's Dreadnought, Harold Jarvis's Marseilles King. In fact, I am certain that a whole issue of The Racing Pigeon could be devoted to the descendants of Osman's Old Billy if sufficient time could be available to research The Racing Pigeon archives.

Wanstead Wonder which won London NR Combine from Banff and London NR Fed from Thurso was another remarkable descendant of Old Billy which, like Forlorn Hope, was bred in 1902. Unfortunately he died young as a result of becoming poisoned through fielding locally. Wanstead Wonder was bred by A H Osman from another daughter of Old Billy when she was mated to a cock bred by the then famed John Wones that was bred from a son of Wones' 1st Bordeaux winner in 1899, when mated to a Clutterbuck hen that undoubtedly was an extremely well bred distance pigeon. Peter Clutterbuck had a great record in long distance racing and won two NRCC races from Lerwick in 1904 and 1914. The Clutterbuck family was founded in 1884 but was dispersed during the First World War, about 1916. A number of them were bought at auction by Sir Thomas Dewar, the Whisky magnate who later became Lord Dewar and whose famous lofts were at East Grinstead and here the Clutterbuck pigeons excelled. Several highly successful fanciers owed much to Peter Clutterbuck who was noted for his enthusiasm for long distance racing and had the distinction of making the first record from Lerwick (Shetland Isles) in 1898!

Peter Clutterbuck would send a great number of pigeons to Lerwick once entering as many as 68 in an NRCC Lerwick race! The Clutterbuck strain was notably dominated by the Belgian strain of Jurion, and they were equally noted for crossing with the Osman family, particularly with the descendants of Old Billy through a daughter of Wanstead Wonder. However, the real point about all this historical references is to try as much as I am able to impress upon fanciers particularly the lesser experienced the power of the genes.

Large commercial breeding studs (not unlike the famous inventor of Coleman's mustard) have proved successful, not only through the number of pigeons, or tins of mustard sold, but owing to the wastefulness of both! Particularly is this true of pigeon fanciers who seem easily pursuaded to part with vast sums of hard earned money in order to win a few short races when the true worth of pigeon breeding and racing must surely be in the production of a family of pigeons that will endure, a family that are capable of consistency in races culminating in success in 500 miles or 600 miles in day of toss races.

Today, more than ever before, fanciers appear to be all too easily persuaded to switch their favours from one family, or strain name as a result of lack of success in the shorter races. Most of which is in the majority of cases owing to lack of patience. This is a great mistake. In years gone by, fanciers generally were not able to, or it seems from my own recollection of bygone days, not so easily pursuaded to switch their loyalties. In most instances, if they did seek a change, it was in the majority of cases, motivated by ambition to succeed in the production of successful day of liberation 500-milers. Gladly I believe it is more than possible those ideas are now once again becoming more predominant in the minds of many. Without patience, and carefully maintained records of breeding, management, and training you can and will not succeed.

Once you have obtained pigeons of known ancestry, and reputation, you can go to work with confidence. Among the pigeons you have are a few that possess the genes for soundness, constitution and reproductive ability. Only hard work, keen observation, and the constant upgrading of your annual breeding will enable you to learn the true worth of your stock. More importantly it will enable you to discover the best reproductive line in your stock.

When you do reach this point of knowledge you will then be in a strong position to line breed to your finest breeder, in the same way that Col Osman line bred to his champion producer, the mealy cock Old Billy. Not only did F W Marriott and Vic Robinson each owe a great deal to the extraordinary reproductive genes of Osman's famous mealy cock but also to the A H Osman policy of line breeding consistently over the years to his mealy No 59 better known in later years as Old Billy. Scottish fanciers achieved great success with the descendants of Old Billy.

A black chequer frill hen bred by A H Osman proved a highly successful breeding hen for J Hendry of Broxburn. She was bred from a son of Forlorn Hope and the Weinberg Hen. Another highly successful breeder for Mr Hendry was mealy cock No 1947 bred in 1919 from The Policeman (777) which was in turn a son of Forlorn Hope and the famous Osman stock hen, known always as 493, which in turn was bred from Wanstead Wonder and the Gainer Hen. Many fine pigeons were bred in Scotland as a result of the Hendry family. My old friend the late Gavin McKelvie was always trying to locate descendants of the Hendry family; unfortunately he died too soon.

In later years and by 1937 the year that John Kirkpatrick with an already well established family won the Scottish National from Rennes with Coronation Express he had become aware of the value of the F W Marriott family. Especially the highly successful line in Marriott's family of St Mark, the son of Marriott's Dreadnought, that was put down in or around the years 1915/16. Thus the Marriott's became firmly established in the John Kirkpatrick family, and therein the line to the celebrated Osman mealy cock Old Billy was further enhanced.

Although John Kirkpatrick's other Scottish National winner of the 1952 Nantes SNFC was bred from a pair of Bricoux, the final sale at auction that was carried out by Bobbie Mayo in Manchester revealed substantially that the Marriott strain was still present according to the sale list of 45 pigeons that realised the then fantastic total sum of £2,075, again a significant intensification of the Old Billy line.

The late Ron Mitcheison of Winchester produced a remarkable hen in his very fine red hen 44.8003 which commenced her winning career as a YB in 1944, culminating after three successive prize-winning Thurso 530-mile races by winning 1st Sect, 7th Open Lerwick (Shetland Isles) NRCC. This very fine hen contained several lines to famous Osman pigeons, including Forlorn Hope and the Weinberg Hen which, among many fine children, produced Hopeful which in turn become the sire of Col A H Osman's very last champion racer, the renowned Olympic which excelled in four successive very hard Thurso races flown in the early 1920s.

Although by no means a pure bred family, despite the many advertisements that described pigeons as pure Osmans, this fact that was never claimed by the founder of the strain but, they were indeed closely line bred to the earlier lines of Old Billy. Whenever crosses were brought in they were not only severely tried, but always mated to the descendants of Old Billy. Then, if successful, mated further into the continuing descendants of Old Billy. Later still the lines of both

Wanstead Wonder and Forlorn Hope were used. The early death of Wanstead Wonder was offset after this loss by the retention of all his sons and daughters, as well as the longevity of Forlorn Hope. If time permitted in making this effort there are a great many more examples of the power of the genes, surviving through the medium of the descendants of Old Billy.

From what I have written, and also referred to I do not want anyone to believe that Old Billy and his famous sons and daughters were unique, but they were truly rare. All I am trying to prove is the sheer transmissional power of the genes. There are of course other great breeders I could write about which have also helped to mould wonderful families through the power of line breeding. All I have really tried to do is to try and encourage fanciers, especially the younger and less experienced the need to obtain descendants of well known breeders or familics.

With patience, and a love for the cultivation of soundly constituted pigeons of good breeding, you will succeed. It is not easy to become a skilled fancier, it does take time, but is not an impossible task. The number of very good fanciers around bears testimony to the power of the genes and the determination to succeed by those fanciers, who have the patience to study their pigeons. This is the way to develop an affinity with your pigeons.

Success is not easy, but very, very satisfying once it is achieved. Cultivate your breeding lines, and hatch every egg possible, once you discover a successful producer, be it a male or a female.

MARCH

Older Breeders — Barley — Spots in the Throat — Memory Lane — Canker Darwin and the Survival of the Fittest

A youngish fancier has enquired whether or not it is wise to breed from pigeons that are over five years old. The answer I gave him was an emphatic yes, but with the added proviso, that only provided the pigeons were really well bred and also consistently reflecting good health and vitality. There have been many excellent pigeons bred from pigeons of seven, eight to 12 years, and even with cocks up to 18 years of age. In fact, older still among the males. Yes, really! An old friend of long-standing has today good racing descendants from a famous racing cock that continued to breed good pigeons until his eighteenth year. The name of his pigeon was Finale, and the owner-breeder the famous long distance specialist S G (Jim) Biss of Brundall. Finale was always a great favourite of mine and I often visited the famous Hillside Lofts during Finale's lifetime. He was a really fine built pigeon with a first class physique, splendid feather qualities and a great personality with a temperament to match. In fact, Finale was really quite a character stamping his authority on the loft without resorting to GBH methods of enforcement.

When he was 15 years old he was listed as the cock of the No 35 pair in the Hillside 1973 catalogue. Bred in 1959, his major wins were 10th Open Fraserburgh London NR Combine, 48th Open Fraserburgh LNRC as a yearling, 73rd Open Thurso LNRC, 9th Sect Lerwick NRCC, 12th Sect Lerwick NRCC. Other wins include 1st Fraserburgh, 2nd Fraserburgh, but above all he was a first class producer. His father was The Inbred Cock, a yearling blue chequer bred from a brother and sister mating, namely Rogersday and his sister The Limit. And for those who are concerned about whether or not one should breed from pigeons above five years old the dam of Finale was the famous Agnes a dark chequer hen which was bred in 1947. Thus an inbred yearling cock mated to a hen in her thirteenth year bred a top class distance racer that was an even greater breeder. For good measure Agnes won 1st Sect Lerwick. Her sire was the Odell cock bred by Mr Odell, a Woodford fancier, whilst the dam of Agnes was the old Bert Donachy hen.

My own experience with regard to breeding from pigeons that are old has also been a happy one. My old favourite, the Logan cock Old K42, bred continuously until his eighteenth year when he died as a result of a flying accident. That year, 1960, he bred two very fine pigeons in Logan Star and Young Duchess and their descendants are still giving a very good account of themselves to this day.

The J B Joel red Logan cock was bred in 1932 at Childwickbury, Herts, and among his many descendants, several of whom became famous in other hands, was the J W Bruton hen Bruton Supreme responsible for a number of outstanding pigeons. Bruton Supreme was not bred until 1946 when her sire was in his fifteenth year. A very good daughter of Bruton Supreme was my breeding hen Perfect Lady which I bred from Bruton Supreme in 1958. She flew Pau and Bordeaux for me but made her name as a breeder. Her descendants are still winning to this day and she bred first class stock until her fifteenth year and lived until she was almost 20 years old. If that is not enough to convince you that you need never to worry about breeding from old pigeons provided, and I repeat the advice, they are strong and healthy and moulting successively well, then I will add further evidence to the above examples by giving you my recollections of several visits I made to the Wood Green lofts of the famous Alf Baker.

I well recall handling the famous Red Admiral when he was 12 years old and even when he was much older. Red Admiral sired Alf's Blue Girl when he was 12 years old and Blue Girl in turn (I believe I'm right) proved to be the mother of Alf Baker's famous hen Quartive. In fact, it is also probable that Red Admiral continued to breed well even into his fourteenth year. In turning back the pages of history another example of longevity and success as a breeder is that extraordinary pigeon named Victor bred and raced by E Reynolds of Clapton, East London, who was matched for a substantial side bet in those days in the 1902 Bordeaux National with the famous J L Baker's Little Wonder and, believe it or not (with apologies to Ripley), these two finished respectively 2nd and 1st Open National, truly a remarkable result to say the least. The point I wish to make is that the 2nd Open prize-winner, Mr Reynolds' Victor, continued to breed until he was 19 years old and bred for Bob McKendrick of Motherwell when 18 years of age a son that went on to breed good pigeons for McKendrick until he too was 18 years old. Longevity, always the sign of a sound constitution, is without any doubt at all worth noting. Ionic, a famous pigeon of the J W Logan strain belonging to Fred Cruickshank of Filey, was another who continued to breed outstanding stock and was doing so at 18 years of age in the loft of the famous big spender fancier Hugh Tillyer (or was it Tilyer?) of Fordingbridge, Hants. This was also the loft wherein the famous La Plume Blanche de Pau was housed (maybe to the end of his days). This world-famed pigeon, bred by Mons Ernest Duray, was certainly there in his fourteenth year. Alf Baker claims that much of his success is due to his policy of not racing his birds beyond five years old and mostly stopping them at four years old, thereby enabling Alf to reproduce from proven racers when they were still quite young. Thus, due to their early redundancy from racing, he made sure he had stock which were sound for both breeding and racing which might not have been possible had he continued to race the ace racers to six and seven years old. It most certainly contributed immeasurably to the future prosperity of the Baker family of racers.

Belgium fanciers too are not opposed to breeding from old and proven pigeons. The great Andre van Bruaene's Old Bull is a perfect example for he bred Van Bruaene's Young Bull when he was 15 years old, and from what I can gather bred again when he was in his sixteenth year. No doubt this will emerge when the story of the Van Bruaene pigeons is told as I am certain it will in due time in the light of the recent sensational money bids. There are many more examples of old pigeons, hens as well as cocks, who have helped directly and through their grandchildren in the production of outstanding racers and breeders extraordinaire that I could name. I believe it likely that I have answered my young fancier well enough. At least I hope so.

Barley feed for racing pigeons
Today there is a greater tendency to feed barley than ever before. Fanciers seek to know if this is wise or against the interests of pigeons for racing. There has always been a great number of fanciers who are totally against its use. Indeed, I expect that despite what I might write, and others too, that many will not even consider for a single moment the use of barley as a feed for pigeons. Today also there is a greater temptation to include the use of barley because of the enormous influence the Belgian fanciers have over the average United Kingdom fancier. Writing from experience I have for many years been a protagonist for the inclusion of barley all year round, the amount

"LONDON PRIDE"
Light Blue Chequer Cock NURP58FWH167. Bred and raced by F W S Hall, 29th Open Lerwick NRCC, 1st Prize London Sect only bird on day. First Prize North Thames Lerwick Club, 1st Prize North Road 5-Bird Specialist Club winning six major trophies and Racing Pigeon Special Performance Tankard.

being governed according to the time of the year. It is cheaper to buy than many cereals although today far more expensive than it once was. However, good quality barley is a very useful grain for pigeons. I know many who will not agree, and equally as many who will shake their heads in complete disapproval. I only use clipped barley and only the very best quality is used. It has a far greater ash content than wheat and ash contains valuable mineral matter and is largely phosphate, and therefore, highly desirable for growing stock.

With the advent of Belgian type depurative mixtures pigeons can now benefit from its inclusion in the daily diet. Some depurative mixtures are far better value than others. If you read the many and varied articles that emerge from Belgium and written by the champions themselves or taped and printed by The RP or the Pictorial you will then realise that much of the criticism levelled against the use of these mixtures is severely jolted when you read as I do with great interest the various loft interviews by eminent Continental journalists who report their on-the-spot interviews with the cracks of Belgium, Holland and Germany. Since barley is the major ingredient in the mixing of a successful depurative its use, therefore, cannot be condemned out of hand, or through prejudice, or on the advice of those who have never used barley.

White spots back of the throat
Recently I have received an appeal from a fancier concerning white spots at the back of the throat of his pigeons. I actually dealt with the subject in years gone by. However, rather than diversify further I will to the best of my recollection deal with the white spots at the back of the throat now. This is for many a seemingly serious matter. It has been written that the cause of this offensive matter is because the subjects affected with white spots are too ardent. Whether or not this is true I cannot be sure, but of this I am certain, I do not think it wise to breed from pigeons that possess these offensive spots. Nor do I for a single moment advise anyone else to breed from pigeons that are infected in this manner. They may not even be dangerous, even so they are for me most unsightly.

Three pigeons were brought to me recently that showed these disfigurations inside the throat. From my own observations, when dealing with other fancier's pigeons, I have long since come to the conclusion that when white spots do appear you may rest assured that if these pigeons are allowed to remain, and continue to be bred from, you will soon have a loft of pigeons that almost certainly without exception manifest this complaint. Personally I am more or less convinced, because of what I have noted, it is a contageous complaint. However mildly dangerous it may be considered by fanciers, whose lofts have, for reasons unknown, produced groups or a family of pigeons that are prone to reproduce subjects that develop white spots at the back of the throat. It has even been written that fanciers have won prizes with pigeons that have been infected with white spot. Personally I have yet to be convinced that this is absolutely correct. In my opinion it is an hereditary disorder, maybe even a malady. It is certainly very difficult to eliminate once these spots establish themselves. Furthermore, and I make no excuse for being repetitive, if you retain pigeons that are infected, and continue to breed from them, and thereby inbreeding the scourge, then you will find it almost impossible to eliminate these unsightly spots

Whenever I have been tempted to make a fresh purchase the very first thing that I examine is the throat and if any pigeon that I show an interest in treveals the slightest evidence of white spots at the back of the throat then I forget all about it, nor for that matter would I consider for a single moment any of its loftmates. Not on your life! I recall a fancier several years ago writing to me on this very subject, I cannot for the life of me recall with certainty how far back it was, although I do believe it was about 15 maybe 20 years ago. This was when a fancier from Cheshire kindly wrote to me concerning white spots following my reference to a serious bout of them. The name of that Cheshire fancier may have been Rowbottom, or even Rowbotham, but he was a fancier of many years experience. His advice was to obtain a box of Puck matches, which I promptly purchased, but since I had no birds to treat for white spot I sent them through the post to fanciers who subsequently sought advice on this subject. The results were somewhat mixed.

In time the fanciers concerned ether lost their pigeons or lost interest in the pigeons they cultivated and owing to lack of success in racing either gave up the sport or made a frest start with a new family or strain. Of this I am certain that as far as I can recall not any one of those fanciers were ever successful at the time they contacted me about white spots and most certainly had not been very successful previously. Puck matches were made by Bryant & May, as are Swan Vesta, but the former I do recall were much longer than Swan Vesta and for that reason were easier to use in dealing with the treatment of white spot. The actual treatment was to wet the pink head of the match and rub it on to each white spot. Repeated rubbings were advised, but under no circumstance should a pigeon being treated be allowed to drink for at least an hour after treatment. I am certain that my enthusiastic correspondent was quite adamant that this treatment would prove successful.

Fortunately I have never actually had to try out the treatment, although quite recently I have tried to purchase Puck matches, but without success. However, Swan Vestas are still made and there is also another match similarly made, manufactured by J John Masters

& Co Ltd, which will probably have the same effect. These are marketed under the name of Winners. It may well be that these spots at the back of the throat are results of a blood condition. Whatever the cause one thing is certain, the Cheshire fancier who wrote to me all those years back was most adamant that the Puck matches worked a miracle cure. I therefore hope that what I have written will not fall upon stony ground. Someone, somewhere, if obliged to try out the suggested remedy, will let us all know the results obtained.

It may well be that the old dog fancier's remedy for keeping dogs in good condition could be the answer to white spots at the back of the throat. This canine health tip was at all times keep a small block of sulpher rock in the dog's water bowl. It is just a thought but probably worth a trial period. There are plenty of good tips around and many old fanciers know of old timers who have gone before them who used them and these have stood the test of time. Therefore, I hope that someone somewhere will take the trouble to either drop me a line or send it to The Racing Pigeon offices with instructions for an early publication. Remember this, if you are in doubt about cleaning and washing out your nest bowls, a tin of Jeyes Fluid will, if used according to instructions, work wonders. Paraffin too is a splendid answer to washing out your loft. It will kill off various kinds of insects etc, including grey mite eggs. It soon dries out and after this wash down you can apply a coating of whitewash, or the modern whitewash, which is an emulsion paint. A can of paraffin hung up in the loft will also help those fanciers with breathing problems but make sure you cover it with a muslin or open gauze to prevent any problems to your pigeons. The fumes will help your breating as those to whom I have passed on this tip have testified. Cannot recall where this well-meaning tip came from, although I do have an idea that our Cheshire friend Mr Rowbotham (or Rowbottom) also passed that one on too. Another useful tip is to save your unwanted sump oil from your car. Many fanciers today carry out their own car servicing. If you paint the undersides of your nest bowls with sump oil you will not be troubled with red mite, who only become red as a result of gorging themselves at nightfall with the blood of your pigeons. Thus your pigeons' vitality is depleted and their night's rest spoilt consequently creating the most detrimental set back to your pigeon's health.

Down memory lane

More than 70 years ago my father Bertie William Hall re-started our loft of racing pigeons. A fancier all his life although from 70 years onwards, he maintained through my own loft a continuing interest in the pigeons. He read The Racing Pigeon from its inception, the very first issue being published on 20 April 1898, the day after my father's twelfth birthday, Primrose Day, and he was still reading The Racing

Pigeon up to a week or so before he died on 9 May 1982, having been a constant reader for 84 years. My father first kept pigeons in 1896, and in the year 1908 used to train his pigeons by a belt driven motor cycle combination, one of the very first to do so in the Walthamstow area. Later on that very machine, my father's pride and joy, was exchanged for the famous Harry Williams complete loft of Putman-based racers, including his Thurso winner Rambler, and many of his best winning pigeons flown by him when Harry lived at Penge. The Putman Belgian strain was highly successful in the UK. The Williams' loft was situated in Northcote Road, Walthamstow, opposite St Michaels Church, and we lived next door to Harry. They were without any doubt a very good family of pigeons.

Harry too was a very clever pigeon fancier, especially in short to middle distance races. A number of the pigeons that were raced at Penge were broken out. As a result of breaking these pigeons out one or two would fly back from Penge and settle on the high roof of St Michaels Church, much to my father's annoyance, not to overlook Mr Williams criticism! I knew this church inside as well as out for my old music teacher, a Miss Gunn, organised the PSA, Pleasant Sunday Afternoon orchestra that used to play at St Michaels and in which I played the violin. Unfortunately the 1914-18 war was to put an end to my father's racing aspirations and the loft was disbanded. Several of the birds were presented to local fanciers including several members of the Walthamstow United FC whose headquarters was the Ringwood Castle. This club was formed the year The Racing Pigeon was first published! Prominent fanciers belonging to that club when we became members in 1922, and highly-skilled too, by virtue of their many successes, were such stalwarts as Lt-Col A H Osman, founder editor of The Racing Pigeon; Walter Hawkes; Jack Leagus, famed for his famous racing cock, a red chequer known as Old 79; Tom Hoskins, the dentist well known for his racing exploits, particularly the remarkable success of his pigeon known as Dollar Princess. Walter Hawkes won the London North Road Combine's Banff in 1923, and my father and I both had the privilege to handle the winner, namely Tame Hawk, a red chequer, a Toft pigeon, a popular strain in those days, soon after victory was assured.

Ted Hyder was another member who won the Thurso Combine that same year, as well as Freddie Brooks who actually won his only race that year, 1923, by winning the London North Road Combine's race from Lerwick! His bird was a red chequer white flighted cock named Elf. What a performance for three members of the same club to win all three Combine races from the three longer distances. Freddie Brooks was the president, with Arthur Davidson, a canny Scot serving astutely as secretary. Other members included Pasty Burke, a very good looking Irishman, who as a young man started life as a trainee butcher working for the same man as my father did, "Sonny" Norwood was responsible for the complete development of the well-known Norwood estate commencing behind Hoe Street, Walthamstow, not far from the old Victoria Palace Music Hall in Hoe Street, where my father as a young lad in early teens stood on the stage having volunteered and had an apple shot off the top of his head by the famous Buffalo Bill Cody who travelled the music halls of the UK in those days following his fame as an American Army Scout during the latter years of the Red Indian problems. Wally Austin was also a member. Wally a High Street, Walthamstow street trader known for good quality fruit especially bananas, was renowned for his prowess as a successful fancier and managed frequently to win the first race of the season, both old birds, young bird and any age without ever training his pigeons other than exercising them around the loft twice daily with flights of upwards of an hour. Just as most of the successful Widowhood fliers carry out today! Wally flew well throughout the programme.

"Survival of the fittest"
It is difficult for the less experienced to appreciate the meaning of the above statement made famous by the English naturalist Charles Darwin (1809-1882) but it most certainly is applicable to the racing pigeon. Only the most strongly constituted, mentally and physically, are the ones to breed from and in turn will help to maintain the strains. If you really want to succeed as a racing pigeon fancier, or as a few do becoming reputable breeders of outstanding pigeons for others, then you must be prepared to cull those that are found wanting and continue this policy without remorse.

A fancier friend of mine phoned concerned about a pair of nestmates he valued because of their ancestry that were inconsistent in returning to the loft from races. He continued by stating that they had had the same treatment as the rest of his team, but were often out for a night, sometimes even longer. He considered they were in good order. This inconsistency had been going on for the past two seasons. My advice was to put them down without a second thought. But, says he, the blood is so valuable. There are far too many pigeons kept for that very reason. What a waste of good corn. These two had been beaten time and again by their team-mates. Now four years old, they had never won a prize or shown any ability to race whatsoever. Such examples can be found in plenty if you search your lofts. Get rid of any that are wasters. Better to keep a few than a loft of never-have-beens. Some pigeons simply are without ability to race. Some that have shown some form are finished at three, or four years old. Some pigeons simply lack a sound constitution. Pigeons can be finished at any age. There is no hard and fast rule on this score. Remember it is survival that you are looking for. The slightest sign of any weakness in the make-up of any pigeon irrespective of how it is bred, or how much the parents cost, such pigeons should not be bred from! Such knowledge must not blind you to the importance of ability and soundness.

On the other hand some pigeons are really quite outstanding. Those that are resistant to stresses and strains are quite obviously endowed with a strong constitution. They show less fatigue after a long and arduous race. These are the sort to breed from. When it comes to long distance racing one has to look for certain qualities. Short distance, and middle distance racing is vastly different to long distance racing. One has to look for the pigeons that never appear tired. These are far fewer than many appear to believe. The champion pigeon is what one has to look for. Such pigeons have to be strong physically, and have the courage to match their strength of mind and body. As with us mere mortals, so with the long distance racing pigeon there are those who have plenty of courage but lack the physical strength to match such courage. Not all pigeons are capable of racing 500 miles, and at 600 miles or more they are even fewer. In the true long distance pigeon their form lasts longer and their strength enables them to benefit from being in-form, and so like Marley Westrop's dual Rome/Faroes racer they are soundly constituted, are tireless, and have remarkable powers of physical recovery. This is really the difference between a true long distance racing pigeon and a medium distance type.

If your aspirations are for short to middle distance racing then you have a far better chance of establishing a family suitable for this type of racing than is possible for long distance. Even so you still have to manage them with considerable skill if you wish to become successful in these fields of competition. Even then as with long distance racing you must always be on the lookout for the outstanding pigeon. They are not easily recognisable. Only racing will reveal the qualities you are looking for.

I do not for a single moment consider that the racing pigeon can reason things out as the human mind is able to do. That is what I mean when I referred above to pigeons that have courage but not the physical strength to match extreme courage. They will continue to fly until they can fly no more. Only the very strong-true-bred-long distance racer can carry out such achievements. Because of lack of reasoning power a courageous pigeon will fly until it is incapable of flying further, and if lacking strength will probably become incapable of flight at 300 miles in a 500-mile race unless it is an ideal day with a helping wind.

The true champion racer is an individualist. This does not imply that the pigeon has the ability to reason things out, it simply means that he has the physical strength and the mental power to concentrate on the task of getting home in the quickest possible time. Even then he or she has to be trained and managed sensibly. All too often proven good pigeons are lost because of the owners' greed or vanity, or both. If you aspire to success in long distance racing then you have always to treat your pigeons with extreme care and consideration when they are young birds and when they are yearlings; this is of the greatest importance. Yearling stock that is well bred and well cared for will prove invaluable at two, three and four years old. It also implies that you must always be considerate. By this I do not suggest that you refrain from culling if you really consider it necessary, you simply have to weigh up the pros and cons with the skill that experience has taught you. Old bird culling must be more strictly carried out than is the case with yearlings and two year olds. If you have the facilities; make way for those lightly raced younger birds as opposed to making room for older birds that are constantly letting you down. If such birds are eliminated you will not be cluttered up with yearlings and two year olds from doubtful parents. It is as simple as that. Remember, you are required to retain those that have survived, or as Charles Darwin would have us believe, think only in terms of "survival of the fittest", that is how Nature works and there can hardly be a better teacher. This is perhaps why the long distance champion is always in the greatest demand. The reason that pigeons like Barcelona Wonder change hands for large sums despite being without a written pedigree. Another reason, as I have written on several occasions is that not all young bird strays are without courage, ability and the strength to match qualities given the chance. However, I am not advocating stray catching.

Probably it is not too "old hat" to remind fanciers of the old school, as well as to inform fanciers of the present generation, that one of the most famous pigeons recorded in the history of the racing pigeon was a stray. She was given the name of Marica. However, unlike the 1986 Barcelona International winner, Marica was never raced but helped enormously to establish the British racing pigeon through the breeding skill of the renowned Yorkshireman Northrop Barker (N Barker) who became a naturalised Belgian, and was the sole agent for the famous J W Logan whose strain is still sought after to this day.

Canker problems
Several fanciers of late have reported outbreaks of canker. Even from those who maintain a first class system of strict hygiene at all times. Canker is always present and given the right conditions is ever ready to strike, and in this respect the type of transportation to the race points we fanciers have to accept is far from being without blame. The complaint is recognisable by the presence of palish yellow-coloured

substances within the mouth (inside the beak) or situated in the throat. This complaint can also develop in the crop but this is very much harder to diagnose when this occurs. Usually the first sign of canker in the crop is when the victim appears to have gone off its food. In fact, in the fairly early stages it is because the complaint takes a hold and causes a stoppage through the advancement of the bacteria upon which trichomonas organisms prosper.

My only qualification to give advice is purely practical, simply that over the years I have experienced the upset of discovering canker among my pigeons. Therefore, fanciers who seek out my advice are receiving merely practical experience based on my own obervations and not information of the kind you would receive from qualified ornithologist or a veterinary surgeon known for his interest in the racing pigeon. From the descriptions offered to me by recent visitors seeking advice it is often a type of canker that attacks the digestive tracts. In such cases this is the type of canker known by the veterinary profession as trichomonas gallinae. In this event the signs are as described in the introduction of this reference to canker through the mouth and throat. When you discover that the complaint has attacked the crop it has then reached a rather more serious stage and, therefore, very much more difficult to sure. In many instances this makes a cure almost impossible. However, if it is a pigeon of extreme value then you can today, if caught in time and with great determination, sometimes bring about a cure. But these kind of efforts for extremely outstanding breeders, or maybe a champion racer, but not always can you be certain of success. However, because they are pigeons of outstanding qualities they seldom appear to contract the complaint. Or for that matter no outstanding breeders and/or racers contract advanced canker! Even so this does not suggest that one should not make it policy to treat your pigeons for canker at least twice a year. This is what the wise fancier does. In this day and age it is a must.

I am not at all in favour of a cure for pigeons that are weak or show the slightest signs of weakness. However, I am fully prepared to try and bring about a cure on a bird that is lesser valued, otherwise I would not have acquired the knowledge in a layman-like manner that has enabled me to cure the complaint when it strikes down a subject of considerable breeding and/or racing value. But you do under such circumstances have to spend much time and show considerable patience. However, bringing about a cure in an advanced case gives one enormous satisfaction.

There is another type of canker which attacks the lower tracts of the digestive system and is seldom diagnosed until it is too late. You often find squeakers in the nest suffering from this form of canker. An examination of the navel will soon reveal trichomonas gallinarum. The navel will reveal a small growth and in a more advanced stage

MARCH

of its development it will be almost impossible to cure. In my opinion it is not worth the trouble, it is best out of its misery. It is not ever likely to develop into a first class racer.

Canker is one of those complaints that you will find is sometimes associated with certain lines in the family tree. Although all pigeons are quite capable of developing canker, strongly constituted pigeons appear to possess built-in antibodies. In a loft where space is considered important for the well-being of the inmates, and a constant vigilance is maintained against the retention of any that show any form of weakness, or ill-health, you will find an almost complete absence of victims of canker. It is for this reason when dealing with squeakers that show signs of disease, especially canker, that I advocate eliminating them from the loft as quickly as possible. Importantly too, I have noted that in many instances fanciers who constantly endeavour to cure the weaklings are seldom successful racing men. Without wishing to appear brutish, to be a top class fancier you must at all times be prepared to cull without consideration of pedigree any that show weakness in their make-up.

In this day and age, more than ever before, eliminate the weaklings without thought or consideration or profit whatsoever. Fanciers who breed pigeons for resale have a moral duty to perform. I once wrote as follows: "Fanciers who are interested in the production of young pigeons for sale purposes have a moral responsibility to their fellow fanciers. Kill and keep killing those that are weak, imperfect or lacking in constitutional vigour". Today more than ever before squeakers are being offered at astronomical prices. I do not decry this so long as the stock offered is absolutely sound. In any case you have the option to buy. You should also be given the facility of rejecting or refusing to accept those found wanting, or suffering from any form of poor health at time of receipt or up to a full week of receipt.

Remember too that although many fanciers appear to consider that any that are offered at a low price may not be any good at all, many fine youngsters purchased at a low price have gone on to reveal themselves as outstanding racers. From what I have read and what I have learned from others in conversation or by letter, a high number of low priced squeakers have in the course of time proved themselves invaluable as racers or breeders, and in some cases both!

Among adult pigeons you will come across pigeons that are trichomonas carriers. In all probability this is the reason that certain lines within a family are known to be producers of pigeons that suffer from canker in one form or another. Yet it is also known by many over the course of time that not always is the nestmate of a canker victim affected by the protozoan parasite. Although the parasite is passed from the crop of one of the parents to the crop of the squeakers. More often than not you will discover that one of a nest pair is completely free of the parasite that causes canker. I have given this matter much thought and have also learned that others have experienced this. It may be the squeaker that is free from canker is the one to be retained specifically for breeding, especially if it is a hen.

Many years ago I recall vividly being offered the choice of one of a pair of squeakers from the nest from a very fine cock that I had persuaded a very dear friend of mine to buy at an auction sale I conducted at the old Clerkenwell Central Club and Institute. The pigeon was one of a very good draft which I sold on behalf of the famous Tom Clarke, one time goalkeeper for Aston Villa, before the turn of the century. He died only a few years ago at the ripe old age of 101 or 102. The selection made for my dear old friend Harry Ashman was an especially robust healthy look cock, and I offered to loan Harry, who lived at Old Hill, my very best stock hen, Lady Barbara. The choice for me was easy — one of the squeakers was suffering from canker, and since no one was allowed into the individual iron barred loft, securely locked with a pucka army issue padlock, 1914-18 type, all feeding and watering being carried out from the outside, the victim of canker had not been noticed. This squeaker was suffering from trichomonas gallinarum, the type that I dread to see since it attacks the lower regions and is very difficult to cure.

The pigeon I took became a splendid bird for me, being a winner as a yearling in a hard race from Berwick, later flying and winning from Rennes. Later still he went on to win prominent prizes from Thurso in a North Road Five-Bird Specialist Club, being 2nd, 5th and 10th prize-winner in successive Thurso races flying this race point four times. His name was Tumbleweed and to prove the constitution of Tumbleweed, a son bred in 1971 raced with distinction for several years as well as proving the sire of good distance pigeons.

APRIL

Buying Pigeons — Winning Hens — The Kestrel — Arthur Sheppard — Colour Breeding — Dordins — Loft Design

By the end of this month we will be well and truly involved in the racing season. The weather is rarely all that pleasant for the start yet nevertheless despite this state of affairs a number of previously successful Widowhood specialists will win their races under this now more popular than ever system of racing. From hearsay and articles in the RP Pictorial it does appear that the majority of UK fanciers will soon be racing on the Widowhood system. For those like me who have enjoyed success with outstanding hens it all seems rather a pity. But that's progress it seems. It therefore pleases me enormously when I learn that a hen has popped up and beaten the crack widowed cocks. They noticeably often do this when the distance is long and arduous!

I keep hoping that those who still race hens and win with them would drop me a line when they do score. It would also be interesting for novices and newcomers alike, and doubtless even the old timers like the writer to be able to read of success in the racing of the hens, including their history (ancestry) and racing preparation. To many it would appear "old fashioned" and outdated. Anyhow I do hope that readers will take the trouble to advise me when the "ladies" beat the "all fired" widowers whose sole object after all is to race home to their "ladies"! Neither do we mere males have to forget the ladies, for you can still race a good hen as others have.

The year when Kelly Jane won Sect F National FC for Joyce & Gerald Stovin in the Colombovac Pau Grand National it thrilled me no end. Again for me it was pleasing to note that a very good inbred racing hen of their old family figured in Kelly Jane's breeding. It was equally pleasing when I read that the winner of 1st Sect and 3rd Open Welsh SR Blue Riband race from Pau that same year flying a distance of more than 600 miles for C Bradshaw snr was a hen, a pencil blue that had previously flown from Pau as a two year old. The parents I am told were bought at an auction sale in the Midlands when a draft of Willy Clerebaut birds were sold. This highly successful Continental strain has been used to considerable advantage by UK fanciers, especially in Wales, and particularly at the distance.

The amazing exploits by Kuypers Bros of Holland with their noted hen De Pau Duiven fill me with admiration. The Kuypers Bros' skill, coupled with the determination and stamina of this now very famous long distance racing hen De Pau Duiven (The Pau Hen) a winner of 1st Open Hens Pau Dutch National, 1st Open Hens International Pau, 2nd Dutch National Pau, 2nd Open International Pau, also 1st Open Hens National Barcelona, 5th Open Hens International Barcelona, 3rd Open Dutch National Barcelona and 9th Open International total Barcelona. What a marvellous series of performances for a hen in a country where the racing of cocks on Widowhood is almost on a par with the numbers who race Widows only in Belgium! However, it is general knowledge the Dutch fanciers are very fond of engaging selected hens in the long distance National and Internationals from Pau and Barcelona.

In 1988 history was made when Clan Queen won the King George V Challenge Cup from Rennes for Allan McDonald of Ireland. In the Irish National FC Clan Queen was the first ever King's Cup winner into Portadown in the 56 years of its competition. Again in the 1988 Pau National race George Burgess of Wraysbury won with Rosemead Abigail, whose sister won 3rd Open Pau National in 1987. This hen in particular also won 20th Open Pau National in 1988. Thus George Burgess who has a brilliant National record, is yet another who owes a great deal to outstanding racing hens. George's article which was published in the 1989 Squills International Year Book makes most interesting reading as well as emphasises how very useful your racing hens can be if you are willing to race hens.

For many it seems that values are all upside down! As a ready wit exclaimed recently: "Are you talking about racing pigeons or racehorses?". Yet not much has changed since I was a youngster that is if you take into consideration what your interests are, whether it be for sprint racing, middle distance racing, 500-mile racing, 600-mile racing, or even beyond these greater distances the outright winning pigeons of these categories have always been in demand. Consequently the progeny of the pigeons that are the best in the above various categories have been more highly valued.

After the direct children the differences are based upon how close the progeny are to the outstanding winning racers, especially from those that can be described as real champions. The greater difference in this day and age in the prices of stock has been brought about through the emergence of professional breeding studs. Added to this commercialisation is the keen competition that is bound to exist between the various studs, which, of course, is based upon the ability to purchase, no matter the cost, the outstanding champion racers at International, National, Combine or Amalgamation level. In very few instances are proven breeders purchased. For various reasons this is not always possible. Quite often the parents of these champions are not for sale at any price. In other instances they have either been lost racing, or are too old to move.

It is much easier to purchase stock direct from champion racers than it is to purchase direct stock from outstanding breeders! Not all champion racers prove themselves as outstanding breeders. History in fact, bears this out, however, the many who are willing to purchase the grandchildren of the champion racers are able to find out the true value of the champions through the breeding results they obtain from the grandchildren of the ace racers. This is, of course, of enormous help to the commercial studs.

The establishment of a successful family of pigeons is apart from good housing, and common sense management, dependent entirely upon one's success in the discovery as soon as one can of dependable breeding lines. Not at all easy. Not by a long chalk. Much depends upon a great deal of record keeping, a lot of patience, and a great deal of intelligence. Essentially stockmanship is a continued policy of testing the stock you breed through the medium of racing and must be maintained. There are more grandchildren of champion racers today for sale than there ever used to be in the days when I was a young fancier. It is obviously far easier to sell the grandchildren of champions. The breeding studs have made this possible, of this there is no doubt at all.

It used to be said that far too many pigeons are bred simply to sell. And that is what it was like. It is still no different. Today you really can purchase the grandchildren of champion racers in the categories that I have listed above, although it is not so easy to purchase the grandchildren of the champion 500-milers, and those champions at 600 miles and beyond, simply due to the shortage of outstanding 500-, 600- and 650- to 750-mile champions. And in this I stress the title of champion in the strictest sense. Such outstanding racers who achieve success in several long distance races are a rare breed. These are the exceptional and the real champions. The even more rare are champion racers that are also remarkably successful breeders.

Years long since gone, when hens were raced a great deal more than they are today there were probably more oustanding long distance ace pigeons to look at and to read about in the loft write-ups, and Squills Year Book articles than you may find now. Mention of Squills reminds me that through the splendid articles, 22 in total, included in the 1990 publication, can be found further evidence of the value of hens especially for long distance racing and young bird Classic races. Several of these articles I have already read again and again. I usually manage to read one or two before I retire around midnight, and there is no doubt that this publication is better than ever.

To lend further support to the value of hens for Classic racing the article by John Duthie of Fife highlights his career with 1st Open Niort Scottish National Flying Club with his Fifer Lass, not much doubt about her sex with a name like that! Of course raced on the Natural system! Tony Twyman describes his methods and ideas entitled Fitness — Form — Fanatics. His honoured Squills invitation to write owing to his great success from Barcelona in the British International Championship Club by winning 1st Barcelona with Little Sarah. No doubt at all about her sex — and a very fine article indeed.

Fred Holbard's article is a splendid in-depth composition with tributes to his father, his father-in-law and his wife Denise after whom young Fred named his 1987 1st Open Morpeth London NR Combine Classic winner. Not much doubt about the sex of that champion young bird winner either, beating 5,683 birds, flying at a hard found 971 yards velocity. The article by David Delea adds further support to the value of hens especially for distance racing, with David winning the coveted King George V Challenge Cup from Pau with a yearling tick pied blue hen named Greenacres Florence, who created a record for the Grand National by becoming the first ever yearling to win this great Classic race. The article by Alun Maull entitled 'Pigeons, A Way of Life', highlights a Dordin hen that won 1st Welsh National for Ray Sheldon of Aberdare, whilst another hen won 1st Open Welsh Combine Morpeth and also won three times a 1st prize Fed for Mr & Mrs Maull & Williams.

Ray Seaton's Squills article describes his methods in preparing two hens for Scottish National FC Rennes races that between them were 16th Open, 42nd Open, 23rd Open and 29th Open Rennes. This particular article is filled with much information, and most certainly pays great tribute to hens that have successfully raced for Ray Seaton, another article that I shall read again and again. G Satterly & son's article highlights their great SMT Bergerac Combine success when they clocked two in one thimble, one a cock and the other a hen. What a great idea for a 1990 mating! This article with loft photographs that will give others a great idea of how to make use of limited space in the garden to the utmost. Note the Equal 1st Combine with a hen.

APRIL

I am delighted to report that quite a number of my readers have reported success with hens. It would be really something if I could report and extend the details of both the breeding and racing of the really outstanding hens. Among those who have taken the trouble to advise me is Tom Watson of Walthamstow whose recent success was with his very best racing hen.

Equally welcome are the reports that I have had of excellent racing cocks. However, at this stage the plum of the lot is a blue chequer cock bred in 1987 by John Bishop of Edmonton and presented to family man John Lynch of Enfield, who flies his pigeons in the name of Mr & Mrs Lynch and son, Robert, who after the race had himself chosen to name their triple winner The Kestrel. This now highly promising racer was tried on the Widowhood system as a yearling, winning several prizes, including three times 2nd prize-winner. This year John decided he would race him on the Natural system, resulting in three successive 1st prizes winning from Doncaster, Wetherby and Northallerton, flown with North London Fed sitting the same pair of solid pot eggs! John Bishop bred this pigeon from a son of his famous Krauth cock Turbo, a winner of 14 races, whilst the dam was bred from a daughter of Turbo's brother. On both sides The Kestrel traces back to a very well known Krauth stock hen namely the Greengage Hen. Doubtless Krauth enthusiasts will know about this good breeding Louella hen. The famous Krauth cock owned by Louella and known worldwide as Black Ebony is well represented in the ancestry of The Kestrel.

Although possibly somewhat premature to write about a pigeon so early in its career this particular pigeon does show great promise. It is also not because of its obviously excellent Krauth ancestry, although doubtless this will interest enthusiasts for the strain. The real reason is that I was first introduced to John Lynch by a young friend, namely Jeff Collins, whom I met when I brought to light his success as a young fancier of 11 years of age after he had won a Newark Challenge Cup in Enfield Excelsior FC. His father Joe Collins was one of about ten or 11 children that lived immediately behind Albert Hickson's Arterial Cafe on the Great Cambridge Road where Albert maintained a very large loft of racers. The young Collins' family soon found out Albert's weak spot! His pigeons on race days! Lemonade and biscuits and other inducements were the price Albert paid to keep the very young Collins' children from playing hectic ball games in their garden on race days!

Ironically, in later years Joe's son Jeff became interested in racing pigeons, and later still joined forces with John Ellis at Forty Hill where they flew with great success. Even later Jeff became a very good steward at my London Auctions. It was then through Jeff that I was introduced to young John Lynch, then a mere lad of 14 or 15. John was a next door neighbour of Jeff's and consequently became hooked on racing pigeons. This was a good many years back now. After doing a one-day stint at the London Auctions as a helper (when Chelsea played away), many years ago now, John gave up keeping pigeons in order to concentrate on his academic studies. His ambition was to become a school teacher and I am happy to say John successfully achieved his goal. Without pigeons I lost touch with the young student. Years later, in late 1986 I believe, John called upon me out of the blue complete with his charming wife Sarah and their seven year old son Robert. It was a great surprise — a very pleasant one too to say the least. At first I could not recall his surname but I did recall that John was an avid Chelsea supporter, and in fact still is. The reason for calling upon me was their son Robert's interest in racing pigeons sparked off by his father's storytelling of his own pigeon racing interest and the keeping of them when he was a young schoolboy. Quite naturally it is therefore for me a most pleasing reason to be able to highlight the ability of a good pigeon to win prizes, be it a Widowhood cock or a happily mated father. In this particular case both, enhanced still further by the many years since I first met young John Lynch.

The late Jim Lammas, friend of John, also played a part in the acquisition of The Kestrel, for it was Jim who told John Bishop of Robert's great disappointment after the loss of the majority of the Lynch family's YBs in 1987. The Kestrel's devotion to his spouse, and those pot eggs, one of these replaced with a live egg that was seven days from hatching on the Sunday evening of the 13th after John, Sarah and Robert paid me a visit, will we all hope be sufficient to excite to action an even greater increase of vital energy than The Kestrel has shown already! On Tuesday evening when I phoned John, he (the pigeon) was sitting 'tight as a drum' with his hen having to remind him that it was time for her incubation duties to commence. You just cannot get them keener than this.

To put the record straight, and because of my interest in good racing hens, I am not against Widowhood racing at all, as some seem to think. If I could get my act together once more I would be very happy to try out the system with a small team of Widowers. Pigeon racing is a physical hobby, young fit pigeons, require younger and fitter handlers than this old codger. Training and management coupled with your own lack of freedom of movement is not so easy when the years have taken their toll. Arthritic bones are a great handicap. However, I still have ambitions to try out the Widowhood system. But would not on any account give up the Natural system by going straight over to this method until I had learnt a great deal more, practically, about the system than I know at present. From what I have read and learnt from others who have discussed the system it does appear that yearlings are very tricky to handle, and therefore would be left out, or used sparingly, and even then treated with great care. Older pigeons, two years, and even three years old would be included. Temperament is of great importance. From what I have read about the best fanciers in the UK and the best on the Continent like our pigeons we too have to learn our trade.

Importantly, theory must be put into practice. To compete against the best exponents you not only have to be prepared to work hard at it, but in order to carry out the work you also have to be in good physical condition yourself because managing and training a team of racing pigeons is very demanding. The sport of pigeon racing is exciting, fascinating, frustrating and often disappointing, but if you love the hobby, and your pigeons, you will overcome the memories of many disappointments. As the famous Arthur Sheppard once wrote: "Nearly 40 years of pigeon flying has left me with the memories of many disappointments, despite the great number of outstanding successes that I have enjoyed. To the young and enthusiastic fancier I would say, stick to your guns, and fight on, if you do your best and work hard you will eventually succeed". Without a doubt Arthur Sheppard established a great reputation for long distance racing.

Often have I thought about my old friend Arthur Sheppard, and wondered how he would have handled the challenge of Widowhood flying. His career commenced before the turn of the century, although it really took off officially in 1901 when he first joined Woodford & District HS and remained a member until the day he died. His son Ron also became a member at a very young age and remained an active member, like his father, also until the day he died. For almost 20 years Arthur Sheppard was president of Woodford & District, a position for which he was nominated by his friend of longstanding, the late Lt-Colonel A H Osman who was also an active member of the club for many years. During his long and indeed illustrious career Arthur Sheppard had the distinction of winning London NR Combine six times. In 1930 Woodford Star won the Combine's Lerwick race, in 1931 Ambition won the Thurso race, in 1932 and 1933 True Blue won both the Combine's Lerwick races. In 1937 Good Lad won 1st London NR Combine Thurso and that same year No Fuss won London NR Combine's Lerwick race, winning a total of seven NHU Gold Medals for Meritorious Performances. Arthur Sheppard also was twice awarded the coveted Championship of London Challenge Bowl in the first five years of the founding of the London Social Circle. Of the many trophies won over the years two others that were equally treasured were the NHU Government Challenge Bowls. Arthur Sheppard created a dynasty. Furthermore the descendants of his famous family are still performing well in long distance events to this day, as the deeds of Ian Benstead and other members of the BICC and EECC can confirm, including those members who had pigeons from Ian.

MONTH BY MONTH — in the loft

Readers may well wonder why I have gone to so much trouble to illuminate the deeds of a fancier long since dead. There are several reasons. Of those I consider important and probably of great interest to many are enumerated. This is what the late Arthur Sheppard quoted when I interviewed him more than 50 years ago:
"(1) It is of the greatest importance that you buy the best pigeons you can; start with a good strain. It is useless obtaining pigeons that do not contain long distance blood.
(2) Retain your foundation blood, and introduce from time to time a really good racer, or a son or a daughter of a good racer.
(3) Try to produce and maintain a medium-sized pigeon that when held in the hand fills the hollowed palm of your hand, not deep keeled but nicely boat-shaped with very tight up-like vent bones, your pigeons when pressed with the firmness of your hand should not give like a concertina.
(4) When holding a bird put the thumbs across the rump and bring the forefingers up under the tail, if the tail gives freely and spread out like a fantail that pigeon will be no use for long distance racing or the production of such stock. Back feathers and rump feathers should continue well down onto the tail.
(5) The wing is important. Flights should be wide, not necessarily long but with quills like whalebone. The cover flights are very important. When the wing is closed the cover flights should be like a carpet and cover well over the back. When a pigeon is standing sideways the cup of the wing should reach down and more or less hide the whole of the body.
(6) Contrary to many fanciers' ideas the tail should be short, certainly not long and loose fitting.
(7) Feed three times a day by hand. Hopper feeding provides the birds with the opportunity to feed at varying times, but your youngsters never seem to get a real good meal.
(8) When exercising your birds use the same tone of voice always and make certain this same tone is maintained when they arrive from the races.
(9) Never use medicines".

Arthur Sheppard never used any form of patent medicines. (Wonder what he would say if he was alive today?—FWSH) Use a small quantity of cooking salt when you are rearing youngsters. A little citrate of iron and quinine, lightly colouring the water was the only mixture he added to the drinking water once they reached the 200-mile stage.

The following will doubtless prove surprising to many fanciers who are interested in long distance-type racing. Arthur Sheppard whom I knew for many years, and visited his beautiful home a number of times. He never fed maize! How about that?

His favourite mixture was good quality maples, at the ratio of five parts peas to one part tares. He was very keen to get good tares although he would be hard pressed to obtain these today. With the addition of a little good quality wheat when they arrived from races and also on Sunday, when they also received a little linseed. He was also advising fanciers to use linseed at least twice a week. The linseed was placed around the loft in clean spots. After the final feed of maples and tares he used to top-up with a mixture of equal parts canary seed, millet, rice and linseed.

Individual training was of great importance to Arthur Sheppard. His YBs when being prepared were eventually given single-up tosses at 45 miles. He believed in this and carried it out religiously. He used to take them in his car, with someone, usually Ron, sometimes in the later years Ted (Porky) Bacon. The routine was to release a single bird at intervals along the road, slowing down at suitable places; you could carry this out in those days! Today the motorways are a different kettle of fish. But I am sure you will realise its importance in creating individuality and independence, and consequently confidence in your young stock. Experience of this kind stood them in good stead in later years. This was Arthur's methods, and I really believe that even today they are not far off the mark for successful racing in long distance racing events. The only observation I would make is the matter of not using maize. Yet in those days quite a number of fanciers who specialised in distance racing were not too keen in the use of maize. Makes you think does it not? Especially today with the advent of Widowhood mixture, especially if you have read that excellent book written by Dave Allen. Note too that Arthur Sheppard never ever mentioned the use of garlic to me, and as far as I know never did use garlic, and he did not use beans or barley. Nonetheless he was always a very hard man to beat, and produced in my opinion one of the finest type of long distance racing pigeons I have ever known. There is a great deal more I could add about Arthur Sheppard but I am afraid it will have to wait for the time being. So I will conclude this memory with a quote the great man once wrote: "The love of the sport brings out the best and it is the best which succeeds".

The Dordins

Many years ago it now seems I was involved with Jim Biss in his idea of establishing a top class stud of Dordins. Through my London auction work I had acquired considerable knowledge of the strain having been called upon to sell several successful Dordin families. At first it was far from easy as little was known about this highly successful racing loft, known as Villa Patience, in France. After all it was about 25 years ago maybe more when I probably sold the first Dordins to be sold at auction in the United Kingdom. At the same time I had gained considerable respect for the Dordins because of their success in keenly contested races against the best lofts in France and in a number of contests the best fanciers of Belgium. Little of this knowledge was circulated among the British fanciers owing I believe largely to the language barrier. Nevertheless a number of English fanciers including Sam Jones of Blackheath and Colonel Hopas, and a canny Scotsman had made it their business to call upon Pierre Dordin at Villa Patience, and make discreet purchases.

Sammy Jones, provided me with the opportunity to sell direct Pierre Dordin imports, and equally gave me the chance to study Dordin's original pedigrees. From the first I was impressed with Dordin's policy of always stating the colours of his pigeons and with special emphasis in the colours of the parents of the actual pigeons supplied. This way one could build up a formidable and useful collection of information about all of Dordin's pigeons. As each Dordin sale promoted by me in London took place, and there were a great number, I made photostat copies of all the pedigrees of Pierre Dordin bred pigeons. As a result I discovered the colours of the parents of Sombrero, who was a mealy cock, and his nestmate brother Sosie. Both exceptional racers and both produced exceptionally well. Sombrero, competed in 31 races, and won prizes in 27 of these races including 2nd Open San Sebastian, 8th International, also 8th Open St Vincent, 8th Open Libourne, and took 35th International in the last two events. Whilst Sosie competed in 29 races, won prizes in 25 of these events including 8th Open San Sebastian, 6th Open St Vincent, flying 1,800 km in ten days and won a prize in each of his last 19 races. Both these pigeons were purchased by S G (Jim) Biss, and besides breeding outstanding stock at Villa Patience, proved very successful as breeders in the skillful hands of Jim Biss.

It was soon after their arrival into England that because of the Dordin information I had already collected that I noted the colour of the parents of Sombrero and Sosie. They were both blues! Pierre Dordin's reply to me was simple and honestle how they, the parents of those famous pair of nestmate cocks Sombrero and Sosie, were mated according to the loft book and so it must remain. When I asked the great French fancier about his opinion on the matter he most certainly gave me the impression that he had no doubt whatsoever concerning the breeding of Sosie! This begs the question what was the relationship of Sosie and Sombrero? They were half-brothers, both were sons of the Bleue de le Crayonne, a yearling paired to the blue cock Radieux, also a yearling. Dordin was a gentleman and a man of highest integrity. I recall only too well despite the passing of the years that Mons Dordin specially stressed to Jim Biss who had shown a great interest in Sombrero that this favourite mealy, namely Sombrero, had that year failed to fill an egg in one of his latest round of eggs. It is now common knowledge among Dordin enthusiasts that

The second loft at the bottom of the garden at Evelyn Street in June 1926. The loft has been decorated with bay leaves, according to Italian custom, for the wedding of Leonora Rispoli, the daughter of the Halls' next door neighbour.

despite this information given by Pierre Dordin of Sombrero's failure to fill one of his eggs, Sombrero was still purchased and this largely owing to Sombrero's international racing performances, for a goodly sum by Jim Biss. Later on Jim purchased Sosie but for only half the price he paid for Sombrero. However, Sombrero bred several good pigeons for Jim Biss. So far as I could see on each of my visits to the Villa Patience lofts no facilities were available for the segregation of any of the inmates including so far as I could judge, or see, any of the ace racers.

Segregation — What it Implies
As I have written before unless you take steps following the first round to protect cross mating through a complete segregation of your pairs, especially the top pigeons in the loft, be they racers or breeders you cannot possibly be certain of the actual sire of each and every egg laid. A controlled first round policy of confinement of each and every pair until eggs were laid as I have outlined is an assurance of the paternity of your first round of eggs, a good enough reason for the value I place on my first round eggs, whereas you can never be absolutely certain of the breeding of any of your pigeons especially in a loft where only Natural racing is practised. Or even in a loft set aside for confined breeding pairs, or prisoner pairs that have a reputation as proven breeders. When one has the good fortune to discover an outstanding stock pair I really advise that they be kept in their own breeding pen.

Breeders whether it be a cock or a hen, or a pair of breeders really are worth their weight in gold, and should be regarded as such. Imagine possessing a pair of breeders such as Glamis Taco and Glamis Scamp a pure-bred pair of Busschaerts that produced a number of outstanding pigeons including two 1sts Open, London NR Combine winners, as well as other prominent Open Combine prize-winning pigeons for their proud owner Brian Haley of Cheshunt, Herts. These were both bred by Regency Lofts and selected as a pair by Paul Smith, and who also bought them back for a substantial sum, on behalf of the stud. Happily they soon settled in their new home and are responsible for a number of first class pigeons. Unhappily Glamis Scamp the hen of the pair was killed when a poplar tree was blown down in a gale force wind and fell through the loft roof. This wonderful Busschaert hen was the sole victim. A tragedy indeed. But fortunately for Paul Smith his wife missed being under the falling tree by the narrowest of margins. Mrs Smith had been busy cleaning out the damaged loft only minutes before the alarming incident occurred!

Having looked at the matter of colours and colour breeding a young fancier who recently called, has been surprised to note that he has produced a blue from a pair of blue chequers. This should have been included under colour breeding controversy, however, since he has been advised by one fancier that this is not possible, I feel it must be dealt with now and without delay. Furthermore, another query has it that a blue from a pair of chequers can only be a hen. The answers to these two queries is that you can breed blues from a pair of chequers, and furthermore of both sexes. It is just possible that fanciers get confused, on such matters, or accept statements that are given in all good faith, but given verbally, mishear and/or become confused or mistaken in the passing of time. Or as some do get hold of the wrong end of the stick from the start. If you pair two blue chequers together you can expect to produce blue cocks and blue chequer hens as well as blue chequer cocks and blue hens. But from two blues you will only breed blues and of both sexes and never blue chequers, or sometimes from two blues you can produce a nestpair of blue hens, or a nestpair of blue cocks. This is emphasised in case a further query is raised! Blacks although few in number appeal to some. Good blacks are not plentiful and I have to admit that I have never kept many blacks, that is self-blacks, except on the odd occasion when I was a great deal younger than I am now. Dark chequers are favourites with me. A colour that does appeal is a black with white flights, similar to Champion Major. This great pigeon Champion Major, a famous racer and breeder of the middle twenties was owned and raced by the equally famous R W Beard of Kenley, Surrey. Before finalising on colour I am also very fond of red chequers, and mealies. Unlike many Continentals I am not prejudiced in this respect.

Canker (Trichomoniasis)
As I have written many times before, the treating of sick pigeons provides little or no dividends at all unless any of the diseases the pigeons are prone to are noted in the very earliest stages. Only recently I noted a young pigeon, a latish bred youngster taken from a first time mating, that did not look as it should. Upon examination I saw that this particular bird had developed canker. Only slight for it was in the earliest stages, a small yellowish cheesy deposit at the back of the throat on the edge of the fringe which could easily be seen by gently opening the beak. At once I gave the youngster one Harkanka capsule, noted the time, and repeated at the same time for the next period of days recommended by the veterinary factors.

In order to make certain I treated the very well looking nestmate, although there was no visual evidence of canker whatsoever, in this case it was probably a waste of time and capsule. These two youngsters are now looking really well and making good progress. Enjoying too whenever possible the glorious sunshine of a very good autumn. Fortunately I had noted the complaint in the early stages. For further precaution I also examined the "foster" parents, the late breds had been reared by a reputable pair of feeders. Both fosters were very well with no evidence whatsoever of disease. They are young in years and have never at any time shown signs of canker sickness or sorrow. It is just therefore possible that the actual parents or one of the parents of the two canker treated late breds is or are carriers of canker. This can and does happen.

Also I have known the youngsters that are free from the disease reared from such a pair perform well or produce outstanding racers. Some years ago a very good fancier friend of mine, alas now departed, had such a pair, yet owing to the fact that from time to time they produced a very good racer my friend of longstanding keenly sought out any that were suspect, immediately putting these down and actually treating the sound ones for the disease. Owing to the reputation of the parents as breeders, they were kept together. Positive arrangements for their eggs to be floated under reputable feeders, were also carried out. Even this did not prevent the odd squeaker developing a mild attack of canker. Yes, there is no doubt that certain pigeons for reasons unknown, except they be carriers of the disease, always seem capable of producing youngsters now and again that show the symptoms. I recall another very good fancier and friend of mine who discovered an expensive imported hen that had proved capable of producing squeakers infected with the protozoa known scientifically as Trichomonas gallinae (canker).

According to various qualified writers particularly the late Dr Leon Whitney, there are several different strains of canker (Trichomoniasis) which of themselves seem to specialise in colonising specific parts of the pigeon (host). There is one that attacks the upper respiratory tract, another the joints and/or muscles (tissues) and yet another that older and more experienced fanciers will be fully aware of, eg the strain of canker that attacks the navel and its proximity. Although not so easy to see if you can catch any of these outbreaks in its very earliest stages you can often cure the victim. However, this is not all that simple to achieve. But whatever happens, do not attempt to save any unless you either catch it out very early, otherwise you will only cause yourself a great deal of work, worry and expense. I am still of the opinion that any that show signs of disease, or weakness should be put down without a second thought, no matter how much they may be fancied or wanted, because of their ancestry!

It is quite plain that Dr Witney went to great lengths to point out that canker was undoubtedly one of the greatest killers among the various pigeon diseases known to mankind. One strain of canker studied, known as the Jones Barn strain, was proven to kill no less than 93% of the pigeons infected. A prolific author, Dr Whitney was also an American veterinary surgeon (a DVM) as well as a member of the pathology department at the Yale University School of Medicine.

The doctor was also a most enthusiastic pigeon fancier. His book entitled Keep Your Pigeons Flying should be included on all fanciers' bookshelves if you can find a copy for it is now out of print. If not at least in your local pigeon club library which if organised properly can prove a profit-making venture for societies prepared to organise a club library. Generally canker will be found in the crops of most pigeons. It is simply waiting for the right conditions, or a sign of constitutional weakness in the pigeon's body in order to make its presence felt. There are several strains of Trichomonas gallinae, some are known to be much more deadly than others. Turkey flocks can be decimated with the strain which infects pigeons, whilst sparrows, especially the English sparrows, can be seriously attacked by yet another strain of canker. Canaries too can become infected with another strain of canker. However, it appears from the work carried out by the various student bodies that the pigeon is the main host.

At one time Emtryl, a product manufactured by M&B (May & Baker) and retailed in convenient packs containing 50 grammes, proved a very useful method of dealing with canker by adding to the drinking water. It was a very convenient method of treating pigeons, especially a large team, without the laborious method of individual treatment which would be necessary if you decided to use any of the capsule or tablet forms of application. Unfortunately the M&B product does not now appear available in sachet packed pouches. For what reason I have not been able to ascertain, possibly economic. Other canker treatment products available and now marketed include Fabry of Belgium who provide two, one capsule (tablet) form, the other liquid, both known as Tricoxine. Again the liquid form is very convenient to enable one to carry out flock treatment, whereas the tablet form is both painstaking and laborious unless you prefer the more positive approach which individual treatment provides.

The Vanhee product, namely Van-Tricho 2000, supplied in tablet form for treatment by dissolving via the drinking water is yet another very useful method of dealing with a serious outbreak of canker. It can also be used as a preventative by flock treatment at the very beginning of the year. However, in fairness to all concerned and the pigeon fancier in particular, the Harkanka capsule and the Fabry Tricoxine tablet, are both very useful for dealing with individual cases. Personally, I find the Harkanka a positive boon for cases such as I mentioned.

However, whatever you may think, or know about diseases within the pigeon loft, only in exceptional circumstances, is it wise to even consider retaining any that show a positive weakness. By all means to your best to learn to recognise the earlier symptoms of various diseases or a lack of well-being in any of your flock. This is all important. Whenever I am in the loft my eyes are everywhere. It is a habit that has developed over the long years and has proved very useful to me.

Today there is far more help given or provided for the Fancy by the various veterinary and manufacturing pharmacies than ever there was in my young day. Many of them have consultant vets who can be phoned. Diseases like "going light", "thinning", "snots" and "yellow spot", usually provided the sum total of known pigeon diseases. Equally amazing too were the methods recommended by well meaning old timers as "positive cures". However, for the most part it was generally "heads off"! A policy that even now is for many still the only way. Overall too this somewhat old fashioned and seemingly drastic treatment may not be such a bad idea after all! However, there are always exceptions. Many years ago before any of the above mentioned drug cures were even thought of, let alone available, if I really valued a pigeon that had developed canker (trichomoniasis) in the throat I always used a mixture of perchloride of iron to which was added a small amount of glycerine to ease the severity of the former. The mixture was painted inside the mouth of the pigeon with the aid of a genuine camel hair brush. If you ever decide to try out this method of treatment make certain you sterilise the brush both before and after use. The pigeon I cured, previously a winner and sire of a winner, was the exception. After being cured it won a well contested Fed race with several thousand birds competing. Later still it was purchased by the famous Captain Ahern for whom the pigeon later bred a champion racer.

However, today fanciers are not obliged to use the methods of the past since the modern methods of curative, or preventative treatments available, even if more expensive, are certainly less painful for the victim than using tweezers or remove cheesy deposits and painting afterwards with perchloride of iron. Such treatment can take upwards of several days, even a fortnight before a positive cure can be affected, as I have known by personal experience, and a long memory. Of this you may be certain that the presence of canker among your pigeons will undoubtedly lower their ability to perform well in the races. Maybe it is for this reason that many of the professional fliers both at home and overseas, particularly the Continental fanciers make a point of treating against trichomoniasis (canker) for old bird racing, young bird racing, and for breeding. They (the Continentals) fly their pigeons for large sums of money in pools, singles, doubles and special prize nominations and leave nothing to chance.

Unquestionably the racing pigeon and indeed all breeds of pigeon, both show racers, and the fancy variety are able to build up an immunity against the protozoa. If this were not true then far more pigeons, especially the racing pigeon for obvious reasons, would succumb to the disease we know as canker. Although as I have written the Continental fanciers appear to favour flock treatment since pigeons are known to be carriers of the disease it may well be that my own efforts to treat individual cases as they appear have saved me a great deal of money which flock treatment would have cost owing largely to the numbers kept! A moral here somewhere! Again I have noted that when a canker infected squeaker does arise the nestmate usually radiates excellent health and possibly developed or born with an inbuilt immunity that has enabled it to successfully oppose the attempts of protozoa invasion. To wit the example quoted earlier of my special pair of late breds.

If you are really observant you will usually find that the parents or rearers of an infected squeaker decline to feed it as soon as they have realised the presence of this diptheric-based disease. Remarkable too and adding more weight to the argument of those who are opposed to the flock treatment method, is the fact that the parents, or rearers, of an infected squeaker rarely ever appear affected by the presence of trichomoniasis (canker) found in those being reared. One thing is certain and according to many qualified bodies and personages canker is mainly a hot weather disease. However, for those who prefer to give their pigeons flock treatment than I am very pleased to report that since I first initiated the idea that I would write about diseases, and canker in particular, I now learn from May & Baker that although Emtryl in sachet packs and as retailed previously has been discontinued you can still obtain through various drug companies Emtryl in 500-gramme containers for around £10. Several fanciers could share this amount and equally the costs, by measuring out 50-gramme quantities. It also is possible that a few trading companies may have a few packs of Emtryl on their shelves although whether or not it has a long shelf life I would not know. One would have to examine the packet, or ask at the counter for such information. Years ago canker was often referred to as "yellow spot" and often canker was mistaken for thrush and vice versa.

Many years ago, before the 1914-18, war most fanciers swore by ordinary writing ink as a means of effecting a cure for canker. Others used a mixture of one part iodine to three parts of glycerine. The writer has used this successfully. About 100 years ago both Lumley and Fulton (1895) recommended one part carbolic acid to eight parts glycerine. But remember this there are different strains of canker and some are more severe by far than others.

You could be lucky and cure a valuable bird that had contracted a milder form of canker, and carry out the same treatment on another to find your patient becoming weaker by the day. Strains that attack the lower regions, the intestines and the tissues at the rear of the crop, the liver, the pancreas, even the heart, are difficult to overcome. Usually student type research has revealed that only a single organ was infected or attacked in a single bird. The experts, that is those who are qualified scientifically to be able to carry out research on the subject of trichomoniasis (canker) have not in total agreed as to whether or not canker is contagious. My own theory based entirely upon observation and experience leads me to the conclusion that for lack of clinical evidence that although it is contagious because most pigeons are carriers and as such are able to develop an inbuilt immunity for a particular strain of canker, they can fall victim to an attack by a strain of canker that is foreign to the resident strain of canker. And not all of these strains are of a mild nature.

Many years ago I recall loaning a very good stock hen to a dear friend of mine, the late Harry Ashman of Oldhill. The hen was Lady Barbara, a remarkable hen that bred a number of outstanding 500-milers. The time came for me to visit Oldhill, Staffs and take my choice of squeakers from the nest. Actually I had little choice and even offered to leave my selection with Harry but being the man he was he would have none of it and so I took 57.575 which eventually won three times from Thurso, also 1st Berwick and without previous training flew Rennes and Guernsey with distinction.

Tumbleweed's nestmate, for that became his name after winning prizes no matter the direction he was sent, was full of canker and the navel was swollen severely. Yet Tumbleweed never bred a canker infected pigeon in his whole life and bred a number of exceptional pigeons. As I have noted personally if fanciers only have the patience to note such matters you will invariably find that whenever they discover a canker infected squeaker in the nest, and the nestmate is uninfected, invariably that squeaker grows up into a sound pigeon that appears never to show sign of sickness and sorrow throughout its whole life! This was the experience I had when the late Alfred Hancock sent me a very well bred pair of East Langton Logans way back in 1943. One was sick with canker and unfortunately owing to a long delay on the railways was almost choking of canker. Attempts to save it with the aid of a severe mixture of half perchloride of iron and half glycerine proved of no avail. The nestmate proved none other than my celebrated Old K42, one of the most brilliant racer-cum-breeders that it was ever my pleasure to own. What is more Old K42 never had a day's illness in his long life of almost 18 years. He actually bred two excellent pigeons. Logan Star and 60.9806 the year he was killed in a flying accident.

As a conclusion I have turned to my much thumbed copy of The Pigeon by Wendell Levi and I quote with acknowledgement the following: "Canker is probably the most widespread and best known of all the diseases of the pigeon. It is an exceptional pigeon breeder who has not had it occur at some time in his loft. Pigeon literature of all countries discuss the disease, but no works noted have described its origin or its cause". However, Levi refers to a number of scientists by name beginning with Rivolta in Italy in 1878 who appears to have been the first to isolate Trichomonas gallinae. Since that time it has been reported from South Africa (Jowett 1907) as Trichomonas Columbae; Hungary (Ratz, 1913); the West Indies (Waterman 1919); Holland (van Heelsbergen, 1925 and Bos 1932); Japan (Oguma, 1931); America (Cauthen, 1934; Waller 1934 and Niemeyer 1939). Levi continues and I quote: "Specimens of throat canker forwarded to Cauthen (1936) were positively identified by him as being caused by the organism and being typical lesions resulting from its presence. Stabler (1938a) has suggested that the proper name for the organism is Trichomonas Gallinea rather than Trichomonas columbae, and again (1938b) that the trichomonas of the turkey (T.diversa) is identical with the organism which is found in the pigeon".

For the record canker can be and has been noted in well managed lofts. If it were not so why do the professionals make a ritual of carrying out flock treatment on an annual basis? However, mostly its presence is more likely to be found where the conditions are far from optimum. Dirty water, especially filthy water pans, or water that is allowed for want of protection to become fouled with excretions, contaminated food, damp interiors to the lofts, as well as a constant lack of sunshine owing often to bad loft design. The ideal conditions in which Trichomonas gallinae flourishes best. Infected corn is always a hazard, and any other lack of even simple rules of hygiene. According to Levi cracked corn whether old or new is a constant source of danger in the spread of canker. So too is contaminated food and grit and poor ventilation a constant source of danger, but to repeat myself (for which I make no apology) especially lack of sunshine.

To sum it all up Trichomonas gallinae finds it much more difficult to survive in a well managed hygienic loft, where floors, perches, nest boxes, water fountains and food containers are all kept clean. Above all else where food is never fed on the floor. This undoubtedly is the safest way to avoid being plagued with the dreaded disease commonly known as canker. As I wrote earlier it really was my true intention to write about other diseases of the racing pigeon but time and space have caught up with my efforts.

Before concluding this chapter I would like to persuade the younger fanciers, that provided your pigeons are bred well and have a good ancestry whatever strain or family they belong it is not easy to lose the ancestry if you have sufficient self-discipline. That is avoid being influenced by the clever advertising that could influence you to scrap your existing stock repeatedly, in exchange for sprinters whose only pedigree is often a meaningless list of ring numbers and very little else. Patience is a virtue and never more virtuous than in the make-up of a racing pigeon fancier. The descendants of the Robinson pigeons are still winning races.

Lastly it is true that far too many fanciers, particularly the inexperienced irrespective of age become easily brainwashed into believing that the winning of long distance races is made very simple by purchasing pigeons that are bred from Belgian-ringed pigeons irrespective of ancestral background. This is both misleading and deceptive. Ancestry, is important at all times, but especially if it is of greatest importance in the production of long distance racing pigeons.

It is equally misleading to suggest as some would have young fanciers believe that the mating of sprint pigeons to old plodders will produce fast flying winning long distance pigeons. Or as so many are led to believe instead of putting fresh pigeons into your stock, that the addition of special vitamins, self-styled secret elixirs and superior made speed cakes is all that is required to help the novice, or the uninitiated and there are many who come within such ranks, or the fancier who has been in the sport several years, but completely without success, or at least has so far not very much to show for his pains in the way of racing success. The real secret of success is to learn to recognise muscular fitness. To such fanciers I would urge them to learn to recognise the difference between a pigeon that is fit to race and one that is not.

This can best be achieved by constant observation as well as the regular, careful handling of your pigeons as they increase their flying power. Fitness for racing is brought about through a succession of flights, that purposefully gradually increase in distance over a gradual period of time. It is the same for the human athlete, the thoroughbred race horse and the racing greyhound. It was also recognised as all important by Vic Robinson and his devoted and dedicated wife Lillie.

MAY

Form and Condition — Educating Babies
Early Widowhood — Pellets — Caring for
Breeders — Going Light — Worms

With the month of May most fanciers' thoughts are directed towards their selections for the June and July classics. The main consideration being "condition", or as some would have it, "form". These phenomena are dependent one upon the other. For the consistently successful both these seemingly magical conditions are recognisable once they emerge. You cannot have form without first producing condition; those who are without experience yet trying so hard to become successful, and have not grasped the true meaning of those seemingly "mysterious" descriptions "condition" and "form" in that order, it is not easy to grasp fully. In fact, it is a subject that has puzzled many fanciers and some seemingly for the major part of their pigeon racing careers. Phenomena are very difficult subjects to understand fully; they are even difficult to explain, or for that matter write about!

Condition can only be the more fully understood and recognised as a result of knowledge obtained through a long period of practical observation. You really have to learn to recognise the emergence of condition in the racing pigeon. Having reached such a stage in your career you will soon begin to notice that as a result of your management, which has brought about a superb racing condition in your racing pigeons, you are on the way to being able to recognise, or maybe a better word estimate, "form" — that magical word so keenly sought by the outstanding racing pigeon fanciers of the day.

To reiterate, condition and form and their eventual revelation are seemingly both magical and mysterious. Once you have learned to recognise these two qualities you will then realise immediately that your management is based entirely upon common sense logic. When this happens you will the more successfully reach the pinnacle of your aspirations. Or if I may be pardoned for the expression, without clouding the issue with highly technical and scientific explanations — you will as a result of your own observations appreciate "condition" when it arrives! When your pigeons look bright and appear happy, looking big, yet when handled feel light yet firm and buoyant, you too have cause to be pleased for you have established a correctly reasoned routine management. Then and only then will you become consistently successful like some you know, and others whom you often read about.

Around 40 years ago when the majority of fanciers within the United Kingdom raced their pigeons on Natural system of racing, that is the method of racing both sexes, or to be more precise paired pigeons, it was much easier to note the emergence of condition in your pigeons. At least that is my own opinion as a result of many years' study. It certainly was the experience of the successful fliers who were overall the most observant, dedicated and hard working. Such fanciers who had set their sights upon a series of races in June and July carefully planned their management to encourage condition to the highest degree. Once their race team had developed racing condition they exploited to the full all that they had observed, even to the point of individually observing the varying degrees of incubation and the consequent mental reaction of each one of a pair.

Owing to experience such fanciers had already learned the need to note the reaction according to the idiosyncrasies of each and every pigeon that constituted the team. Each pigeon must be treated as an individual, for each have their own likes and dislikes. Through the achievement of racing condition which encompasses racing weight, superior muscle, both firmness and condition, each one of these phenomena combines to bring about a mental and physical co-ordination, and these highly important facilities of both mental power and physical power, within mated pigeons, or Widowhood pigeons, at the right time, lay the foundations for the development of form. What in fact are the important signs in a healthy pigeon that signify or are symptomatically associated by the experienced fancier with form?

First, it is important for me to emphasise that condition is not brought about by some magic elixir or secret formula. There is no prescription whatever for good health. A weakling is a weakling, and will never become an athlete, for that is really how I look upon a sound healthy strongly constituted racing pigeon. Cod liver oil, used over generously on corn, or whole groats, pinhead oatmeal or wheat is a great mistake. Used sparingly it is of enormous help and is a great boon. When a pigeon is in good condition for racing there is a most definite brilliance in the feathering. The eyes are brilliantly bright and dry, the eye cere should be a dry chalky white, and this chalky whiteness should also be reflected in the wattle. The skin on the breast should be clean and completely without scurf, and show a healthy pink colour, even a deep pink in the case of Widowhood cocks.

On the sternum (the keel or breast bone) as I have mentioned many times over the years you will in a "form" pigeon note a small red mark. A dot if you like on the sternum is a certain sign that "form" has arrived. The inside of the throat should be a healthy looking light pink. The feathering all over should fit tightly, especially the neck feathers should shine in a scintillating style without any being out of place. Wings tight fitting to the body, with a richness of colour in the feather. The muscles of the breast feel firm like a brand new tennis ball. Condition is paramount and can be produced through good routine daily management. Form is the greater enhancement of condition. As with racehorses, condition is obtained through regular routine work, indicative of the Vincent O'Briens, Henry Cecils and Michael Stoutes of the equine bloodstock world who produce condition of mind and body, upon which "form" is based. Healthy racing pigeons, like healthy racehorses are happy creatures, and if this is produced with consistency, you have the right formula for success. Provided you have much patience, and profit from what you observe and also possess pigeons of good breeding and sound physique you cannot really fail to succeed. Again a final reminder. You simply can't rush racing pigeons into condition. It is achieved through painstaking routine daily management. Efficiency is regarded as a science. If you know your job as a pigeon fancier and only race your pigeons the Natural way (on the Naturally mated system) you will always beat a second rate self-styled Widowhood expert. You will at times even beat the crack Widowhood fliers simply because the crack Widowhood man cannot and does not have pigeons on form every week of the racing season through to the longer distances. The crack Natural racer will make certain that his pairs are not all mated on the same day, but in weekly and/or fortnightly groups, even monthly groups. Armed with the added advantage of good racing hens that will always be hard to beat, especially at the distance — what better way to improve a family of racers.

Conversely the Widowhood specialist who really knows the job inside out, and can also claim stock sense as well as stock superiority will also be difficult to beat. Remember also that weather plays an enormous part when one is flying Widowhood. Weather is always the great master over all. It is equally important when one is flying the Natural system. However, the great advantage when the weather is difficult, lies with the clever fancier who knows the value of a good hen, and at the same time maintains a team of Widowhood cocks, and a team of Natural pairs of racers. The best of two worlds!

Whatever system you practise the pigeons must be brought into condition. It is then that the observant fancier will recognise the outward signs that signify in certain pigeons that form is about to emerge. One's ability to do just that is the key to successful racing. When a pigeon is in "form" it can tackle successfully three, sometimes four successive races. A good pigeon with good ancestry whatever system you favour, in form, can last around 21 days. This does of course favour the cocks, and thus glorifies the Widowhood system where they can stay in condition for many more weeks. Remember this too, all pigeons have different reactions when in form. You really have to get to know your pigeons individually. Again I emphasise that once you are able to get your pigeons fit (reach racing "condition") they will be happy creatures.

When the reigning chaser/hurdler trainer Martin Pipe was interviewed by Richard Pitman up at his stables, Mr Pipe who was about to overtake Michael Dickinson's all-time record of 120 winners, and may even train a staggering total of 135 winners, stated somewhat ominously, "Once horses are fit and kept happy, they are able to take several races in a short time". This is equally true about racing pigeons. It most certainly applies to pigeons on Widowhood, and of course Naturally flown cocks. Hens flown under the Natural system for obvious reasons (egg laying) are not quite so easily handled. However, with care and impeccable date records of egg laying, you can look forward to enormous success with outstandingly keen hens. They are simply quite incredible, as are all females — regarding toughness and tenacity.

What really matters, most of all, is whatever a fancier can get out of a good trier, the utmost effort, be it cock or hen. Only the most observant will succeed. In the case of those raced on the Natural system the keen fancier notices everything that is important. The same applies to those who specialise in flying their pigeons on Widowhood system. With careful observation you note down any pigeon that shows a particular interest in the making of a fresh nest in a completely new site. A few feathers, the odd twig or two, anything in fact (except fine string or cotton which is quite dangerous by becoming entwined around the ankle or foot), then whatever you do for heaven's sake allow the nest building to continue. Life in its various forms on planet earth is really a matter of two basic principles, ie procreation and survival. These two most outstanding faculties of life for man are no less important for the racing pigeon. This most fascinating of creatures will strive to procreate and survive, regardless of the difficulties it may encounter, or be called upon to face up to. Shortage of food of a regular nature and they (the racing pigeons) will

MAY

very speedily adapt its diet to whatever is available. Despite many difficulties, even injury and some forms of sickness, they struggle hard to overcome all hardship — they survive!

Equally they will then do all in their power to procreate themselves. If the sexes are imprisoned there will be no difficulty in their reproduction. But should the sexes be kept separated, and as I know from experiments and years of experience, you provide lath partitions down to floor level, with space between each lath, wide enough to permit eggs to be pulled through, as well as the luxury of deep litter (although dried earth and a mixture of powdered droppings will do just as well!), cocks will soon pair amongst themselves and take to eggs from a group of four eggs laid by a love mating of hens — so powerful is the instinct to procreate. Pigeons suffering from an attack of paramyxovirus will also do all in their power to survive and procreate. Even among mated pairs infertile eggs which often are the legacy of paramyxovirus do not prevent them from continuing to procreate (by laying eggs) until the anti-bodies have built up efficiently for their eggs eventually to become fertile once again!

To those seeking advice because their babies are loathe to fly away from, or even around the loft, do not worry too much, or lose patience, but east in the wind does not help at all. Let them out at regular times but not with a full crop. You have to educate your babies to respond to your call, whatever form that may take, be it a coaxing "come on, come on", whistle, or corn in the tin. If you are able to separate the first round from the second round of young ones, then do so. Often despite the petulance of east winds since March 21 when the sun crossed the line the separating of the first round from the second round will often motivate the former to fly, instead of encouraging the earlier bred ones to lay about, or idle themselves on the loft roof. However, my own personal view is it is wiser to wait until after June 21 when the sun crosses the line again. It may then be westerly and that will bring about a complete change in the sky and cloud formation. By waiting until then you are the more likely to have your entire young bird team still intact.

We have now been long enough into the season's racing to note as well as to learn from many of their successes as well as hear from those who have been unsuccessful. Therefore, by this time of the season fanciers wherever they may live, or whatever route they may fly their pigeons, if they have been unsuccessful to date will naturally wonder where they may have gone wrong. Often it is very difficult to understand what is actually wrong with their management. Especially is this so when after questioning these unsuccessful fanciers one is quite unable to offer a solution to their problems mainly concerning their management they appear to have done everything that is possible! Mostly I consider lack of success is because of a combination of feeding and training. Either too little, but probably too much of the former, and too little of the latter! In a few instances it could be too much training, and too little food, coupled with a complete inability to recognise racing condition. This is largely because of lack of experience in the fancier. Feeding and training when mastered will solve many problems for the unsuccessful.

The present trend in management is towards the Widowhood system made famous some 60-70 years ago by Renier Gurnay of Verviers, and in the UK by Fred Shaw, Billie Pearson, Arthur Wright and O I Wood. From my own reading of the many interesting articles which appear consistently in the monthly Racing Pigeon Pictorial it is fairly obvious to me that the majority of the successful fanciers who now fly their pigeons solely on the Widowhood system (viz racing cocks only), were previously successful when they flew their pigeons solely on the Natural system (flying cocks and hens mated and according to nesting conditions). This upon reflection is most noticeable. In plainer terms many of today's successful Widowhood fliers within the United Kingdom were counted among the most successful when flying their pigeons as mated couples (the true Natural system) and thus enjoyed the successes that good racing hens bring about, a feature that is denied them once they go over entirely to the Belgian method of true Widowhood.

There are, of course, exceptions to the rule, as one notes when reading of both successful UK exponents as well as the Continentals who in a few instances favour the inclusion of a good hen or two for the International marathon races. However, these are exceptional. Most of the crack Widowhood fanciers stick religiously to the racing of cocks only. The point of this observation is that from the experience of many of whom I know, and the many more that through the excellent Racing Pigeon Pictorial articles I have read and re-read is that the majority of the successful fanciers flying Widowhood had mastered the art of conditioning pigeons for racing on the Natural system before tackling the latest fashion for racing in the United Kingdom. Remember it has taken the British fanciers many, many years before the present numbers took to the Belgian system of Widowhood racing as being the best way of racing. Many there are who consider it is the most successful system ever.

Yet despite the present trend there still remain some who excel racing the Natural way only! Whilst a few more are beginning to accept Widowhood, others still are inclined to keep a few Natural pairs and thereby have a few good hens for racing. Personally if I ever get round to racing seriously again, and I most certainly want to, I would prefer to keep both systems going! The racing of hens has always appealed to me. Furthermore I enjoy the atmosphere of a well-managed Natural racing loft, and the added thrill of seeing my pigeons enjoying their surroundings with freedom and liberty which is denied in the rigid Widowhood loft. If one has enough space I believe it is possible to enjoy both systems. Fortunately I have enjoyed the pleasure and success that good hens can achieve when well-managed. It is, therefore, pleasing to note the success that even a few Continental fanciers enjoy by the inclusion of a few well bred hens. However, it is also true to state these are becoming more rare in this day and age.

I would like to draw attention to the many outstanding successes gained with hens by the now famous Emile Denys. This fancier has truly highlighted the possibility of winning an International with a hen and with great emphasis, by beating many crack Belgians and their ace male racers with his hens in the marathon-type International Classics. So much so that I learn that several of the best in Belgium, France and especially Holland are now including a few hens for such events. Of course, there is no doubt at all that commercially the successful Widowhood cock, particularly an International winner has great appeal, and consequently is highly valued. With such high sums being asked for 1st prize International winners from Barcelona, and/or Pau, it is not small wonder that the buyers have to consider ways and means of producing as many youngsters as possible and as soon as they can, in order to safeguard their investment. It is obviously far easier to carry out such a programme of reproduction if you pruchase a cock bird.

For those fanciers who have an outstanding racing cock they too, like the buyers of International winners, can protect their future racing interests by planning likewise to produce as many youngsters as they possibly can by the mating of their very best bred hens to their champion racing cock. It is a sound enough policy. Your ace racing male can be mated to several hens. It does, of course, require careful planning. One has to be fully prepared to use several pairs of feeders. You don't have to build a special loft but it would help if you were able to convert a section in order to carry out such a scheme. I have known of crack racing cocks being mated to as many as ten different hens in a single season, furthermore with enormous success. It means work, methodical planning and painstakingly kept records. This is a sound policy to provide your best male racer with the opportunity to fertilise many eggs in a single season. It is a positive approach to reproducing your very own winning genes within your own loft and you cannot do better than that. It takes time and thought but is simple enough when you sit down and plan it out.

The feeding and care of the stud cock goes without question. The physical well-being of your champion racer, and his mental happiness is of paramount importance. If you are interested in this progressive form of reproduction then initially you can commence with four nest boxes at floor level. Take your four selected best bred hens, and include where possible any physically suitable hen that has flown well, or shown intelligence in her races and place in the nest boxes. Work numerically, according to numerical ring order with the lowest ring number in the first nest box from left to right — you will find this useful advice. Then allow your most outstanding male racer to enter the arena. After a while you will release a hen — try to work it so that you release the hen in No 1 box first, followed by the others in numerical nest box order. It really does help if you work methodically in this manner. Make certain you have a suitable loft book, the loft book pages in Squills Year Book are cheap and very useful, and record all ring numbers of hens in full. Note everything that you consider worthwhile. Within a matter of four hours the cock will have been introduced to all four hens, and from my own experience several years ago, when I experimented very successfully with my champion breeder and racer Old K42, will have trod each hen. Now you can see why I advise that you look after your champion stud cock with great care!

Good sound corn is essential, I have always favoured a wide variety of corn. A good seed mix in discreet quantities is useful for all pigeons concerned, including in this the pairs destined as foster parents to take the eggs from the selected hens that have been covered by your champion pigeons. On the day you arrange for the selected hens to be introduced to the champion cock you also pair together four pairs of what you consider are your best feeder pairs. They should of course be kept in another compartment, but failing this, then if you must confine these feeder pairs to the very same section then make absolutely certain you keep them in the top row of nest boxes. The special hens should be screened off, and the stud cock removed, when a top row pair are released. However, in my own case I managed to use a separate loft, a small one, but very convenient in the carrying out of such an experiment. It also eliminates the complications observed above when special hens and feeders occupy the same section. My record achievement with Old K42 was to produce 24 eggs in a single season of which one pair proved infertile, later in the year I was able to prove the hen responsible for the clear eggs was the cause — she had developed a tumour which later proved fatal. She was an oldish hen, and proves the point I advise, select known producers of fertile eggs, from females that are youngish in years. Avoid if you can inexperienced late breds and yearlings. Go for two and three year old hens with good temperaments.

The late Walter Hawkes, my mentor in pigeon matters, used this method of breeding with both his famous racer 648 and later Tame Hawk, his Banff Combine winner in 1923 (or was it 1924?) that was a direct descendant of the then famed Toft strain. From 648 Walter bred 18 the first year he tried the system. Later on in 1927 having taken on the management of the famous East Grinstead lofts of Lord

MAY

Dewar (also of Scotch Whisky fame), he tried out this form of breeding with a famous Logan cock that his Lordship purchased at the final J W Logan sale. The result proved successful and most pleasing to all concerned. But again I must warn you it is not for those who are not fully prepared to acknowledge the work involved and the intense dedication such a breeding plan entails. It really does pay to safeguard the eggs from a first class racing cock, or even a first rate racing hen, for these are the most precious form of reproductive jewels you can hope to obtain, and thereby worth all the effort such a venture demands.

Pigeon racing is a matter of survival of the fittest and there really is no finer way of maintaining such a policy than progeny testing. This is what you would be carrying out by adopting such a policy. Breeding from your absolute best racer mated to selected hens, and these in turn subjected to training and racing in a most vigorous and testing manner. Remember it must be only the fittest that survive. Better still, and a definite step further in selective breeding would be where possible to mate your champion racer to as many of his grand-daughters as is possible. Failing this then make certain that from those you produce in your initial steps to breed from your champion and his maybe even unrelated, but selected mates, that you take steps to mate him at a later date to his grand-daughters that will eventually emerge as a result of your special matings. Only success will follow if you go about it with great care, great patience, and with all the enthusiasm of the dedicated. That is one of the safest and quickest ways in which to produce a champion racer. If by good fortune you already have a grand-daughter of your champion cock, then include her among your selected hens provided she is old enough to breed from. This would be an instance where you ignore my advice to use only two and three year olds. But despite this latitude, if your only grand-daughter is too young, then forget using her until she is old enough, even if it means waiting another year.

By this time of the year your old birds that comprise the race teams should be looking the part and show by their very actions to be what I call "quick and lively"! By this I mean alert, ready for a fly whenever they are given the chance, all this despite the many who still race their pigeons Naturally and have their racers paired, feeding young, sitting or driving to nest. Pigeons that are feeding will quite naturally not be quite so keen to fly for long periods, or interested to leave the nest box for too long a period of time, unless they are without feed for their young or missing grit or minerals so essentially important.

Fanciers going over to the Widowhood system of racing have found that owing to these changes of system there are many at this present moment of time anxiously settling their first round of young birds to their outside surroundings! The wise fancier gives a great deal of attention to the safe settling of their first rounders taken from the first matings of their Widowhood cocks since, for many, these will be the only "round" likely to be taken from these Widower cocks until racing is over! It follows, therefore, that every possible care must be taken to hold these babies. Foolish acts can quite easily lose squeakers, try to remember that too many losses will cause a great upset to your future plans.

Not everyone can boast the ownership of a few proven pairs of good class breeders, well bred pigeons retained because of their outstanding ancestry as producers to assist in the replacement of early losses. Successful pigeon racing requires good team work between the owner/manager and his/her pigeons. Good team work can only be obtained as a result of practice. Pigeons, like many other kinds of animals, and even other birds, can be encouraged to react to good habits. You as manager are responsible for teaching your pigeons good habits. Pigeons themselves are ideal subjects for the training required in order to encourage them to react to your "routine" demands without any form of cruelty whatsoever. Their naturally innate impulse enables the patient and keenly observant fancier to train his young pigeons to respond to his commands, according to the environment and situation and style of the loft. But without any question at all it will only be constant practice that will bring this about. Practice makes perfect for it establishes discipline. Without this quality success is undermined.

The owner/breeder must carry out the plan, and provided it is a sensible plan, he or she will still have to take all logical steps to see that the plans are carried out to the utmost perfection but again I would reiterate only a consistency of practice will enable your young pigeons to learn and remember all that you teach them in these early days. Leave nothing whatsoever to chance. To train young pigeons in the manner they should be trained and "routinely" encouraged to become habitually responsive requires a great deal of patience and a lot of practice. to quote from an old but well-screened TV advert: As the dear old lady was told when she enquired from a road worker "How do I get to Carnegie Hall?", the reply from the well-meaning road worker was, "With a lot of practice"! So despite the fact that we are months away from young bird racing it will pay you to always think in advance and take every opportunity to control young pigeons by ensuring they become "practice perfect". This applies to all sports and pastimes but never more so than when dealing with racing pigeons especially babies. Fanciers often say to me "my birds won't eat the maize", or others say "my birds won't eat beans"! Others will state "my birds waste a great deal of corn by scattering!". The answer to all three is "you are overfeeding", or "you are hopper feeding with far too much mixture available to them".

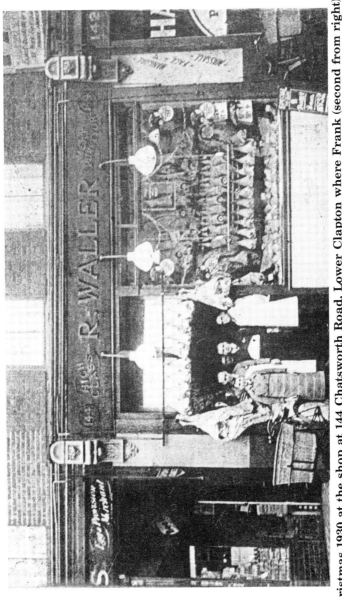

Christmas 1930 at the shop at 144 Chatsworth Road, Lower Clapton where Frank (second from right) had been responsible for dressing the window. Frank and his father, who had sold out to Mrs Waller, raced to a loft behind these premises in the Clapton & District FC. They also raced in the North East London Wednesday club of which Col A H Osman was a founder member and who had successfully proposed Frank for the secretary's job.

This is not the way to feed racing pigeons or control your pigeons. If you really want control you will have to feed them accordingly, and this does not imply, or means a starvation diet. Not at all. That is downright cruelty, there is no need for such measures. A long time fancier whom I have known for several years called recently and was another who was surprised by my reply when he told me his birds would not eat the maize contained in a very economically-priced mixture. The maise was actually dogtooth. Again I replied, "You are giving your pigeons far too much to eat!". When I showed him what an ounce of corn really looked like he surprisingly observed, "that looks like a starvation amount", or words to that effect. So I spent some time explaining to him what he must do, even if it was only carried out for a few days so that he could note the difference in his pigeons' activities.

I also told him to find his copy of Squills 1989 and read and re-read again the article entitled "Chalk and Cheese" written by The Red Baron, otherwise known as The Megaphone. That article alone is worth the price you pay for this ever popular book. At least it is if you thirst for success on a consistent basis. Neither of the fanciers referred to in this most informative article overfeed their pigeons. The two systems described by The Red Baron were a sharp contrast of styles yet they both keep their pigeons "quick and lively"! Good trapping is of paramount importance and pigeons that grow fat and lazy through over-indulgence and lack of exercise will never be "quick and lively", they will become a team of slow racers and bad trappers.

My friend who bought the mixture containing the dogtooth maize has since called again and I quote his observation: "They don't leave any corn now, everything put before them has gone, no waste at all, marvellous". Once you take the trouble to work out an improved controlled system of feeding you will then learn to recognise fitness, and also witness a better response to your commands and not waste a single grain of corn, you do not have to starve your pigeons. After a time you can even afford to reward your birds by handing out a small amount of titbits. It can comprise a simple condition seed or any one of the advertised conditioners. You can even use a feed supplement and there are enough of those about these days. Or you can make up your own with a mixture of pinhead oatmeal and whole groat base, a fifth of each, into which is added a fifth linseed, a fifth rice (paddy rice), a fifth canary seed. Mix these five ingredients thoroughly, then add some cod liver oil. Don't overdo this, sufficient to coat the mixture, then dust it or "dry it off" might be better words to use, with beer yeast powder. The mixture can be used the next day.

On the question of pellets as a complete daily feed I am still not fully convinced. However, today you have a far greater variety of pellets to choose from than ever before. Most certainly those now manufactured, especially for racing pigeons, can be used as part feed. I have used them all in turn but I still prefer good old-fashioned corn provided it is sound and clean and of good quality. Maybe I would feel happier if I knew exactly what each brand of pellet comprised; the ingredients and the percentage would help me to decide in the same way I am able to select mixtures of the good brands of mixes. There is one pellet that has recently become available that I believe could prove useful after a hard race or a bad training flight. Pellets are probably very useful to use after long enduring flights, they digest more quickly than hard corn, and are ideal for tired or exhausted pigeons and when rearing. Whatever you decide to use when the final decision or policy is finalised remember for racing up to and including 470 to 500 miles that one ounce per bird is the basic requirement needed to which according to work and effort on the part of the pigeon you can add, in the same way as those two fanciers described by Red Baron. Once you discipline yourself along these lines you will see a great deal of difference in your pigeons and learn to recognise when it is necessary to build a daily intake of say one ounce plus an eighth of an ounce, or even more. Remember not all pigeons weigh the same. The fit pigeon sparkles and looks big, yet when handled feels light and buoyant, but firm like a new tennis ball. Warm feet, bright eyes, with tight fitting silky feathers, white wattles, dry ceres and clean feet.

Feeding is an art, breeding is an art, training is an art, management is an art, mating pigeons is an art. Yet all these skills can be learnt. There is no dearth of good reading today a great deal more than there ever used to be. If you want to succeed all you require is the patience, a great deal of observation, a lot of patience and willingness to change your methods, even your pigeons, especially if you have only attained modest success. Unfortunately, far too many are ready to condemn their pigeons without a fair trial. In latter years due to problems of health, the difficulties of getting about as well as other moral obligations I have found it impossible to race pigeons as often as I would wish. For the last few years OB racing has not been possible, but from time to time a little YB racing has been carried out.

Paramyxovirus has been one of my problems, vaccination had not been carried out whatsoever, when in mid-September 1987 I was caught out with a severe outbreak. It was by then pointless to subject them to a vaccination programme. At one time I had as many as 68 severe cases, each victim contained in single unit isolation boxes. In due time I discovered if you have the patience and desire to give every victim of paramyxovirus a chance to survive, with care there is no need to put any pigeons down. Even the worst cases could be saved! For at least six months all my pigeons were given only water that had been previously boiled. Each isolation unit had its own water vessel and its own feed receptacle but the intensity of the work proved enormous.

The most stubborn case of all was a 1982 bred cock. He was struck down with paramyxovirus at the beginning of October 1987 and was unable to control his shaking head movement. Until February I was beginning to wonder if I would succeed. His moult took longer than normal to complete — especially the end primaries, I then decided to use beer yeast tablets. After a week my patient with two tablets a day was showing a great improvement, the dose was increased to four per day. It must have been pure coincidence but the continued improvement was quite astonishing. This pigeon is so improved he was soon calling to nest and ready to pair at the first opportunity.

The paramyxovirus is not all that my pigeons and I have been subjected to. But this will have to do for now since my real purpose is to answer several fanciers who have been greatly worried as to whether or not their pigeons that have contracted the complaint will ever get over it and if they do will their reproductive functions be effected, so I must, in all honesty, continue to relate my own experience. In the YB section where my 1987 bred young birds still remained after the moult, the 1988 bred birds, including several that had raced quite well, had mated among themselves and commenced to produce a colony. Every day throughout the year and whenever possible they were allowed their liberty. One chequer cock took a long time to get over paramyxovirus, his crooked neck was painful to observe, but he was determined to get over it. His will-power was, I am certain, motivated by a chequer white flight hen. They laid and the first pair more or less dominated my relatively small YB race team winning five trophies and four 1sts in two clubs and a goodly amount towards the corn bill. Without boasting, they are very well bred indeed.

The cock straightened his neck, believe it or not, by feeding my winning pair of young birds. Probably his presence in the young bird section enabled the other occupants as well as the young birds to build up antibodies. I have also used a great amount of Jeyes Fluid mixed with the right amount of water to drench my gumboots by stepping into baths when going from loft to loft. It has proved an enormous task but I am glad to report none of the other lofts or sections suffered from paramyxovirus. However, this is all about this for now except to add, do not keep more pigeons than you can find the time to look after them as they should be, as you will finish up like me, all work and no time to play!

Jaded appetites — Going light

Another query that arises quite frequently concerns jaded appetites, consequent loss of weight, and watery droppings. This often arises among early young birds especially between five and seven weeks old.

MAY

So many approach me on this subject, particularly novices and the uninitiated about this vexing question and I feel moved enough to comment upon it, again but only according to my own observations! Remember, as I often remind readers, that I am not a qualified veterinary surgeon, and regretfully most certainly not an academic ornithologist — simply a very enthusiastic observer of my favourite subject, the racing pigeon, that's all! In an instance of jaded appetite I feel the crop and if it contains an abnormal amount of water with little or no evidence of food treat immediately for coccidiosis. Try as much as possible to encourage your babies to feed by tempting them with tasty small type corns (no beans or barley for this treatment) mixed with a little tasty seed, Red Band, or any of the small seeds, including rice. But before doing so make absolutely certain to remove all water from their section. After an hour or two provide the squeakers with Coxoid through the drinking water at the rate of one liquid ounce to one gallon (precisely) of clean water. Remember too that precise measurements are of the utmost importance. If you only have a few babies then you can do as I do, make up a half gallon by reducing the coxoid to half a liquid ounce, viz four pints instead of eight pints. Continue this treatment for seven days so if it is coccidiosis, you will soon note a decided improvement by the fourth day. If after a week you do not find any improvement then the possibility is that your babies are infected with worms, and more likely these are hairworms (capillaria).

Worms and Treatment
In the event of worms there are several types of treatment available today. One of the most simple and cheapest to use is Harkers New Roundworm Treatment which is used in the water fountain and if used correctly will eliminate both hairworm as well as roundworm. However, I do recommend treatment every eight weeks. Other products in either tablet or capsule form include Ascapilla which is a capsule and is administered orally one capsule to each bird on successive days and although more expensive is certainly most effective. However, as with all orally administered products is also labour intensive especially if you house a large number of pigeons. Another excellent treatment is Spartakon which again has to be administered individually. It is in tablet form and supplied in boxes of 50 sufficient to treat 50 pigeons individually, it is also reasonably priced. One only to each pigeon sufices for a single treatment. Although adult pigeons appear to possess an enormous degree of tolerance in respect of worms it is not so with young birds. In any event worms will seriously reduce a pigeon's ability to perform well racing-wise, whether they be young birds or adult birds. Therefore, treatment is important, but even more so is that fanciers should see to it that their pigeons are kept free

from these debilitating creatures by preventative means. There is a great deal more that can be written about worms in pigeons but time is as usual against me, and I really want to satisfy the many who have written to me concerning bare patches, or loss of feathering particularly on the breast and other parts viz wings, back of neck and even the rump.

Bare Patches

As I have already mentioned above other correspondence includes a fancier troubled about loss of feathers and consequential bare patches on the breast, neck and even the back of his pigeons. All these bare patches are the work of the depluming mite. To those fanciers who have treated their infected birds with feather rot cream, those who can obtain any now can make up your own feather rot cream using one part flowers of sulpher powder to four parts lard and mix well, store in screw-type jar, apply to bare patches and repeat seven days later. However, do not become despondent simply because the bare patches remain. These unsightly featherless patches will only produce a new growth of cover after the moult has taken place.

One-Eyed Cold

Correspondence concerning that vexing trouble one-eyed cold continues to arrive on my desk. It is really quite surprising how many fanciers seem troubled with this vexed matter of inflamation of the eye. In severe cases the eye becomes almost closed. But this is brought about through sheer neglect. Invariably you will only find a single bird at a time and it is most certainly not difficult to detect, for the poor creature nearly always stands out from the rest. But if you neglect to take steps to effect a cure, the bacteria which brings about this complaint will spread to other occupants in the loft or section. The disease which appears to affect pigeons only as opposed to poultry can also in more serious cases affect both eyes.

It is very painful indeed and you should quickly seek help from your local veterinary surgeon who may, provided you take the bird along to him, prescribe coryzium capsules or eye drops. Yes surprised as you may be about coryzium capsules these are still manufactured. Indeed only very recently a friend of mine was not only prescribed coryzium capsules for his pigeon but at the same time supplied from the vet's stock shelves with a tube of these most effective capsules. But you must be prepared to pay for a visit as well as the capsules. You also must be fully prepared to take the patient along to his surgery for examination. This is general practice. In fact, it is the law as applied to the Veterinary profession.

There are other effective antibiotics manufactured. These mainly are produced for the treatments of dogs and cats. They are, however, also very useful in dealing with one-eyed cold in the pigeon. The one I obtain from my own vet for use with the pigeons is namely Neobiotic Eye-Ear Drops. It is contained in plastic type 5ml mini-bottles that enable you to squeeze gently one or two drops only, into the eye affected. My own experience and observation suggests three times a day to be the right allocation of drops. The liquid must be well shaken before use with the screw top safely secured after use. At all times it should be stored at room temperature, and especially keep from freezing.

There is another similar type eye-ear drops made I believe by the Glaxo company, namely Betsolan, this too I have used with equal success. Neither products are really expensive but you must be prepared to pay for veterinary services when seeking these products. Try to purchase two mini-bottles, they invariably carry a good shelf life. Fanciers write to me too, seeking to know if Optrex will cure one eyed-cold. I have tried Optrex but without a great deal of success. However, when you discover a pigeon with one-eyed cold, and if at the time it is not possible to get assistance from a vet immediately, or for a day or so, then Optrex will most certainly provide relief in helping to soothe against the pain caused by eye inflammation. With one other exception there is not really a great deal more that one can do when you run up against these circumstances.

The one exception that I have used to help alleviate pain and suffering as well as cure one-eyed cold (and it is very cheap to make-up) is a solution of one part boracic crystals to one hundred parts boiled water; in plain a one percent solution. Years ago I used to make up equal parts boracic crystals to equal parts hot water. Since then I have found this to be far too severe causing pain and discomfort to the patient. You can apply the solution with the aid of an eye dropper, or by saturating a small corner of lint, or good quality cotton wool and squeezing the solution into the eye making certain you well saturate the eyeball with the lotion. Do this at least four or five times a day. It will effect a positive cure but it will take several days, sometimes in severe cases upwards of a couple of weeks to bring about an absolute cure. But it does work if you have the time and the patience!

Whatever you use for the eye try, if you can, to use the various solutions advised at room temperature. As I have already noted for those who are willing to try out the boracic crystals solution you can make this up for a few pence. Bottle the solution into a screw-top bottle, secure tightly, keep at room temperature, but remember to label the bottle and date it to, stating the nature of its contents. It will remain effective for 12 months. If you keep it by you it will help you out should you find yourself in a situation as outlined above. It is both soothing as well as healing.

Colour Breeding Controversy
Now I wish to reply in particular to Arthur Keegan, who claimed in his interesting article that I had stated "two red chequers cannot produce a blue". This was never stated by me. What I did state was that two red chequers paired together will produce blue hens but never blue cocks and I stand by this statement. Red chequer cocks and red chequer hens, will produce red chequer cocks, red chequer hens, mealy cocks, mealy hens, blue chequer hens, blue hens, note four colours in the female, and only two colours in males. This same phenomena applies to reds mated together (that is reds that lack the chequering associated with the normal red chequered pigeons), these two will never produce blue cocks, or blue chequer cocks, or any cocks that are of these colours including pencils, slaties, or any pattern of chequering on a blue background. The same applies to the mating of two mealies, you produce blues that are always female.

From the mating of a mealy cock to a red chequer hen you will produce red chequer cocks and mealy cocks, and no other colours in males. The hens from this colour pairing will be red chequers, but you will also produce blues and blue chequers from this mating. The same rule applies when you mate a red chequer cock to a mealy hen, viz the cocks from such a mating will be either mealies or red chequers. The hens will be either blue or blue chequer, with the odd dark blue chequer. Never will you breed cocks that are blues, or blue chequers from this colour mating. So you can safely mate the following colours safe in the knowledge that any blues or blue chequers they produce will be hens. The pairings to note are as follows: mealy cock x mealy hen, red chequer cock x red chequer hen, mealy cock x red chequer hen, red chequer hen x mealy hen, remember all blues and the blue chequers and the odd dark chequer will always be hens.

JUNE

The Greater Distances — Dealing with Failure — Young Bird Curriculum Judging Breeders — International Barcelona Hens

This is the time of the year when most if not all fanciers' thoughts and ambitions are concerned competitively, or at least are consciously aware of the great long distance racing events now pending. On the south route we are aware of such race points as Pau, Barcelona, Palamos, Perpignan, Dax, Lourdes. We are also aware of the great Nantes National and Barcelona or Palamos races into Scotland. And never overlook the long distance races promoted by Up North Combine. Then we have the many and various North Road races to be flown from Lerwick in Shetland, and the even larger number of races to be flown from Thurso, the furthermost point in Scotland for the English and the brave fanciers of South Wales. In all these races, you need a special kind of pigeon. They have to be physically as well as mentally strong in order to overcome the varying degrees of difficulty which the terrain of the British Isles provides. And never overlook the varying degrees of weather that we can always expect! It is truly an exciting time for the real long distance racing enthusiast. What greatness awaits those who have prepared their entries with painstaking dedication. Only the very best bred will succeed, and not even them if they have not been properly prepared. Success in the exciting long distance events in the next six to eight weeks can and will only be achieved by the very finest bred pigeons that have been fully prepared. Sending the "old plodders" in this day and age is no use at all!

Simply to enter a pigeon in the vain hope of turning a slow plodder into a long distance champion racer is just a fantasy. There have been many fine long distance racing achievements by Scottish, Irish, Welsh and English fanciers in races from 600 to over 1,000 miles, particularly during the last 30 to 40 years. A review of record performances listed annually in Squills Year Book reveals many outstanding performances. That brilliant day of toss performance achieved by the famous A R Hill of St Just, Cornwall, in the race promoted by the British Berlin FC in 1952 when Mr Hill's splendid red chequer hen flew from Hanover, a distance of 686 miles, and recorded a velocity of 1309 ypm and was ever afterwards referred to as the Hanover Hen. The performance of R P Eggleston's entry in the 1967 Barcelona International negotiating the 938 miles in three days, two hours and 50 minutes to return a velocity of 511 ypm flying to its home in Long Marton, Westmorland, was indeed some performance. It certainly caught the imagination, and no doubt inspired others to compete in extreme distance events in later years. The performance of Alan Raeside's red chequer cock, in the 1975 British Barcelona Club's race flown from Palamos recording a velocity of 675 ypm over a distance of 1,013 miles into Irvine, Ayrshire, was an even more amazing performance, and one that for sheer racing ability will take a lot of effort by both fancier and pigeon to eclipse. But pigeons of this calibre are hard to find. Nonetheless this does not mean that it is impossible to produce a genius which is what they really are. With the best known preparation in the world it will require a brave and courageous pigeon sent in the highest possible state of physical fitness to emulate the Conqueror's performance. Such a pigeon must have all the qualifications that one always associated with a champion.

However, despite the abilities of the truly outstanding long distance racing champions that history has shown can and do appear from time to time with so many wonderful races ahead of us, we may yet see the record of the Conqueror beaten. Nearly 62 years ago that famous Scottish fancier John Brennan fired the imagination of many fanciers through the outstanding performances of his black chequer cock Stormy Petrel that flew San Sebastian, Spain, to 49 Dundonald Road, Kilmarnock, a distance of 862 miles, in three successive years. Stormy Petrel was a truly great-hearted, supremely strong and very determined pigeon. His performances from the one-time famous Spanish race point were an outstanding achievement, and great credit was owed to John Brennan, who was one of Scotland's best long distance enthusiasts.

Failure to succeed
Many fanciers at this time of the year contact me because they have failed to win a prize. Often the lack of success is owing to their own

JUNE

lack of patience to understand more fully the need to produce pigeons that are brimming with good health and vitality. All to often I believe this state of affairs can be attributed to overfeeding. It is also owing to undertraining. If you are still unable to succeed ask yourself and consider in all honesty, the true answers. If the replies you receive are to the effect that your pigeons are consuming too much corn, and not flying freely then you will not win at all. Also at this time of the year one hears of the many and varied methods employed by fanciers who are eager for success, or having failed to succeed, are ready to suddenly change their form of management. This is both upsetting and indeed frustrating for the pigeons themselves. There are many who having set their stall out during the winter months by preparing their loft for a changeover to the Continental system of Widowhood racing, and now without any prizes to show for their pains and outlay immediately condemn the system, and revert to the Natural method without even giving thought to their own contribution in the application of their management. This I am afraid is rather foolish. Perhaps the best advice I can advance for those who contemplate doing a similar change over for next year's racing is to make certain you have the time to attend to the intricacies of the Widowhood system for there is not the slightest doubt that it does take time. It is far better to master the Natural system. If you are unsuccessful with the much tried and tested Natural system then you had better improve your knowledge before you spend a great deal of time and money altering your loft and purchasing expensive Continental pigeons and as many do, purchase an expensive designed loft as well. To be successful at pigeon racing at either of these two systems of racing you really do have to learn the trade yourself.

Impatience is all too often the real cause of failure. Impatient to learn the art of managing a team of Natural racers is the downfall of many, who believe the Widowhood system will bring immediate success. The greatest fault as I see it, portrayed all too often by the novice and newcomer alike, young or old, is their failure to appreciate the need for extreme patience in the beginning. Secondly, there is a serious need to understand that young pigeons must be taught their trade. That is to say they must be trained well once they have learned their home location. They will only accomplish this after many hours and days of flying daily around the loft. This too must be carried out intelligently. For the novice it is hard for him or her to appreciate what this implies. Since fortunately there is always a possibility of newcomers, however remote it may appear to be among our ranks, it is always wise I think to bring this matter up at this time of the year. Particularly for the uninitiated.

Young bird curriculum

The study of educating your young pigeons is of the utmost importance. Whatever your ambition may be with regard to the rest of this season's old bird racing, for those with limited numbers or who have only more or less recently entered the sport, you still have to be aware of the need to look after your current year's stock, There is a great need to make it a golden rule to pay heed always to the weather whenever you contemplate releasing your pigeons for a flight, even when only releasing them for a flight around the loft. Yes, even for home exercise it is a necessary precaution. Do not carry anything out whereby your young pigeons will suffer as a result of your own thoughtlessness.

Already I have learned of enormous losses of young pigeons. It is happening all over again; mostly too they are lost around the loft. When the sky is without a single cloud, and the sun is shining brilliantly, you can lose young pigeons so easily. There are far more young pigeons lost as a result of a thoughtless release from the loft than from any other cause. A liberation of baby pigeons whether it be around the loft, or a trip down the road, can always produce losses if you do not heed my advice about clear skies and strong sunlight or grey skies and no sun. Broken cloud is always a far safer way, and a far, far safer rule to remember at all times when dealing with the liberation of pigeons, especially young pigeons. East winds will also take their toll. "Wind in the east 'tis neither fit for man or beast".

Your pigeons will like you have to learn a trade and so far as you are concerned, their trade in order to bring you pleasure is learning to know their way around the loft, and around and over the local topography with the least possible threat of danger. Try to remember this; they will not acquire this knowledge if they are given their liberty in all their innocence amidst frightening conditions.

Judging the breeders

A young fancier as well as a fancier of longstanding both recently enquired on this very subject. This is a complex and difficult subject for the inexperienced to grasp. Indeed it is also not the easiest of subjects even for the longstanding fancier to fully understand. However, judging the value of a breeder as a result of the success of a particular bird's progeny is the soundest way of advancing or accumulating such valued information. This is known as progeny testing. Testing the young produced from a pair, or a bird or either sex that you have discovered consistently produces stock that are hard to lose, or that produce birds that score in the races, qualifies these producing types as potential breeders. They have what I describe, as well as many fanciers that have gone before, and have shown from past recorded history to be clever breeders of racing pigeons, as producers or proven stock birds. I have also described them frequently as "go on breeders" as opposed to "stop end breeders". The former are those that are annually producing reliable racing stock, and whose children continue to reproduce equally reliable racing or breeding stock. On the other hand the "stop end breeders" are those whose progeny are easily lost, and consequently are not represented at all when the annual inventory is taken.

My own experience is that you are much more likely to produce reliable racing stock from the mating of line-bred pigeons, or even inbred pigeons than you are by the willy-nilly policy of completely unrelated pigeons. Often history has shown that the bringing together of two different inbred strains will produce a much higher proportion of successful winners and breeders as opposed to the crossing of single crosses of unrelated pigeons. Although on the credit side of the crossing of unrelated pigeons, you do now and again obtain something rather exceptional.

The real trouble emerges when you mate these first crosses. All too often the second crosses can be a complete failure. If you do persist in such a policy you can find yourself in possession of the most diverse, and odd types possible or that you would wish for. Provided you are severe in your selection and prepared to eliminate the weak, the misfits, and the incompetent you will be able to both inbreed, or line-breed with impunity, as many of the greatest breeders of racing pigeons have proved beyond all doubt.

International Barcelona Hens
The remarkable hen that in 1986 not only won 1st Barcelona International beating 18,075 competing birds, but was also 18th International Barcelona in 1984 beating 13,015 birds, an incredible dual performance, rightly named Barcelona Wonder, was but a stray. As I have written before and often, not all stray babies or young birds are without good ancestry or even lack ability. The two go together. The two strains that produced the 1986 Barcelona winner are two of the oldest strains in Belgium, Bricoux and Fabry; both are still highly rated in Europe. The former is especially favoured in what used to be called Eastern Europe, particularly Romania. Perhaps it is the swing of the pendulum back to Natural racing that of the 25 Internationals flown during the past five seasons three have been won by hens. Raymond Hermes of Germany, purchased Barcelona Wonder, a German bred pigeon, for a huge sum despite the fact that the only information of her breeding was to the effect that she was a cross of Bricoux and Fabry.

One thing is certain that whilst good hens will emerge from time to time in these Barcelona Internationals they will probably, it seems, remain in the minority so far as being actual outright winners of the

the Barcelona Internationals. At least recent years have proved the success of hens to date despite the small number on average competing in the Barcelona Internationals I often wondered if the Continentals and also the British realise what they may be missing by denying their hens the opportunity in this "equal sex, opportunity, age". Good racing hens, like outstanding racehorse mares, both flat and chasers, often prove to be outstanding producers, which is not so consistently true of highly successful thoroughbred males. There is always that infernal problem of parentage. In the world of pigeon racing, at least you do know the mother of the champion, but not always the father.

Young birds for novices
Here we are once again in June! The longest day will come and go and next month young bird racing begins. Newcomers and novices are all agog at the idea of racing their very own bred "racers" for the very first time. It is thrilling, and it can also be very exciting for the entire family, especially if you have allocated various young birds to members of your family so that they are able to take a more than passing interest in the race in the knowledge that they each have their very own representative. Why not? It is a nice family idea. Let each member of the family give his or her pigeon a name. No matter how frivolous the christenings may appear to be.

Young bird racing fills one with high hopes. Fortunately for the newcomers and lesser experienced, and the novice who has yet to break his novice status (win his or her very first red card), experience does not always count. Often it is great enthusiasm to succeed that counts for successful young bird racing by the uninitiated. In the first year nothing whatever will have been too much trouble for the beginner. All the attention possible is devoted to the "babies", and it really does pay off once they have become accustomed to flying for long periods of time in and around their home, they will be hardening their muscles, and building up their stamina. The newcomer to racing will have really made a great fuss of his or her baby racers giving them their almost whole attention, whilst the established fanciers and old timers will have been very busy dealing with their old birds. To this extent the young birds for this season will have been somewhat neglected. This is often where the novice scores over the old timer! This too is the reason that so often the novice makes a mockery of the word experience. Young bird racing is a great leveller. It is also a most informative method of evaluating your first year's breeding efforts.

Young bird racing is really and truly the novice's racing programme. Beginners and old timers toe the line together. Not always do novices win, and equally not always do old timers win! Mind you whatever your status racing-wise, one will always experience the disappointment of losing a "fancied" pigeon. However, probably the best way to avoid disappointments is not to select "favourites", although this is easier said than done! However, if you provide your young birds with the opportunity of plenty of training tosses in the first 20 miles it will go a long way towards helping them gain both confidence and experience. This method will help you to win prizes, and even a first prize or two! Yes, really. Tosses all the way from one mile up to 20 miles with plenty of intermediate stops en route will pay huge dividends. Better still if you can find the time and the patience to give a few liberations in twos or threes. There are "many roads leading to Rome", or more plainly, many varying methods of training pigeons, especially young pigeons. Teach them well and they will, if they have the right mentality, learn a great deal. This is often the reason that a novice will beat the old timers. Young pigeons need a great deal of attention, and consequently that requires patience which often the old timers are not always prepared to extend. But not so the very successful "old timers"! They know only too well the need for the utmost care and patience, coupled with the most astute powers of observation so necessary to beat allcomers

It is also most noticeable that the more time you spend fussing around your youngsters the tamer they will become. It is always a pleasure to handle manageable young, they also make tame old birds. Remember this to your great advantage in the years ahead. Although young birds require plenty of food until they have cast their first major (primary) flight they must not be overfed in such a way that will make them stubborn when speed of entry into the loft should be carried out instantly upon the word, or call of command of the handler. Quick trapping is of the utmost importance and you must at all times work to encourage this important act of obedience. Although the are problems facing the novice and less experienced, probably the major problem for most is how far should one send the youngsters during their first year? A good question too, and one that requires a sensible answer. A great deal depends upon your future ambitions but there is not the slightest doubt at all that the question poses a considerable variation in answers. Even 80 years ago controversy raged on this very subject. Many were of the opinion that you should race them out to the bitter end. One school of thought is all in favour of racing all young birds out to the maximum distance of the race programme. Another school of fanciers is inclined to favour a very restricted racing programme, sending only to the bitter end those they like least of all. Others still advocate that the most promising should be stopped once you have reached 100 miles. Not a bad idea for those who are just beginning the formation of their first loft.

However, in my own opinion a great deal depends upon a young fancier's future ambitions. On the one hand a newcomer may be interested solely in becoming a good club flier. And why not, there is a great of fun and pleasure in club racing. You also have the opportunity of competing at Fed level. Much will depend, of course, on the type of young bird racing you meet up with. Not by a long chalk is the weather dependable enough to be certain of 100% returns although it is more likely that if you have taught your young birds to fly in small groups they are less likely to be easily lost. You cannot beat the idea of instilling confidence into your babies at an early age hence small group training as I have suggested many times before. Thus if you experience a hard season weather-wise then you will sustain few losses. You may be quite certain that the weather plays an enormous part in the end result, and this is the reason I consider that you should train your babies well. Make a point of dividing them into two teams so that you have a team in reserve, once you have reached the 100-mile races. I have long since come to the conclusion that the education of young pigeons, both racing and training, is of great benefit in the establishment of your family, and two teams are a safer way of becoming established. Among those youngsters of this year's breeding may emerge a real champion, equally if you are even more fortunate you may discover a breeder in the early years ahead from those you bred this year's crop from

Outstanding racers and extraordinary breeders are both very hard to come by. Believe me I really do know after a life-time in the game. You ought really to play safe at all times, for you can never be absolutely certain about the weather conditions. If the weather is "easy going", and if your young bird team has been well schooled, you can take greater liberties by sending a third of your total team to the longest race on the programme. The most important aspect to keep in mind is to avoid total and complete elimination of your first round of youngsters in a very bad race. And this can and does happen, but not if you split your first year's team into at least two teams. To young fanciers just beginning, such a calamity will prove most disappointing — even disastrous. My advice to newcomers is to avoid at all costs those who advise that they start their babies off at 30 and/or 40 miles. Remember, it takes a great deal of time to learn to evaluate weather conditions. Long distance initial training can be very dangerous for the lesser experienced and there is a great more usefulness in training in easy stages. It is safer, and certainly far less hazardous, than an initial 30- or 40-mile toss.

Young bird racing requires a great deal of routine management. Educating your babies to survive in the crates (or baskets as many organisations I note still use) is of the utmost importance, you must avoid fighting at all costs. When you place your "babies" into the crate or basket for the very first time it can prove to be a most frightening experience. Give them plenty of wood chippings, cut hay or peat and plenty of room. Hang the thoroughly clean water trough empty on the outside. Add an ounce of corn for every five pigeons, and an hour or so later, arrange for the empty basket trough to be filled with clean water and trickle it in slowly — this is highly important. This is the hallmark of good training and management. Far too many young pigeons go to the initial races without having been given the opportunity to experience this essential education.

You cannot commence this curriculum too soon. Nor for that matter maintain it for too long. However, personally I find that it pays to "school your babies" (for that is all they really are) by not allowing them to drink in the loft after their evening meal. You then pick them off their perches very carefully and place them into the crate or basket, according to type of container you use, and then place the water troughs in overnight. They will soon learn to put their heads through the bars and drink. Keep this up for a week at least. Thereafter a refresher course once every week. For those with a reasonable amount of young bird racers it will not prove to be an impossible task, but it will prove a most satisfactory accomplishment once you are able to witness your "babies" satisfying their thirsts.

The longest races in the old bird calendar are a great thrill for those with long distance old bird aspirations to the fore. Especially those with established families, but for those without a team of old birds, or adult birds with experience enough for the 500-mile events, the young bird events are now the real target for such enthusiasts, and newcomers. Unfortunately far too many enter the sport and then depart all too soon. It therefore may be as well to pinpoint a few of the problems that face the lesser experienced and particularly those that have only recently entered the ranks. To these fanciers in particular I would extend a warm welcome and hope that they will stay long enough to become indoctrinated into the mysteries and fascination of the racing pigeon. Far too many leave the sport before they are fully able to enjoy the thrill of clocking a winner.

For some time now I have given serious thought to this "state of affairs". All too often the newcomer becomes confused with what is best to do. How best should he or she go about the establishment of their very first foundation family of pigeons? A good question and one that in all seriousness should not, and cannot be dismissed lightly. It is so easy for the experienced to dismiss the newcomer or novice with a degree of impatience, when care and consideration should be a much more suitable approach. One must bear in mind that without newcomers to the sport, within a few years there would not be sufficient numbers to warrant the investment by those who provide a weekly press, let alone monthlies. The right kind of publicity for the sport of pigeon racing is of great importance. Clock manufacturers, loft makers, manufacturers of loft equipment, housing or loft requisites. Not also to overlook the enormous variety of corn grown and mixed. All this and more that is annually invested amounts to an enormous sum. A dwindling membership of the various unions should be of grave concern to us all. We are all, each and everyone of us dependent upon each other, and the sport of pigeon racing in particular is highly dependent upon numbers. Numbers of fanciers I mean.

Without newcomers the sport of pigeon racing would decline to such a degree that those interested in investment would lose interest. Without these business types all those items I have mentioned might not be available or might become even much more expensive! Thereby one of the most fascinating hobbies in the world could dwindle alarmingly. So we "oldies" all have a great responsibility to do our utmost to avoid such happenings. It may at this stage of the year appear somewhat premature to write about the problems that face those about to enter the world of the racing pigeon. It is a great problem at first especially for those who have no one in the family to help or guide them.

One of the hardest tasks — indeed a great challenge — is to commence a loft of pigeons with young birds only. Especially those who are having to take extra care to hold their youngsters that have been purchased from a breeding stud, since it seems that quite a number of fanciers, either from premeditated choice, or through circumstance, start a loft off in this manner. Such fanciers are in the difficult position of being without flying-out stock whatsoever. This is a great handicap but not impossible. In this event they have to be very, very careful with their newly-acquired squeakers which may not all be of the same age. From what I hear, and gather from newcomers that call upon me, there are quite a number of fanciers in this very position. For these fanciers, severely handicapped as they are through lack of flying-out stock, much patience is required. Even this rare quality in mankind will not always help unless you know how to apply such forebearance, as well as being given the right kind of advice to begin with.

Many of the lofts designed appear to lack the facility of a look-out from the front, or a bay affording the occupants with a rooftop view bay that at the appropriate time, such as (in my mind's eye) I now envisaged many are now in need. A simple facility such as I have outlined will enable those with newly-acquired youngsters to place a look-out pen, even a simple training crate so that the newly-acquired purchases can "look out" from within, and develop what I have always termed "spot location". It is quite surprising what young pigeons can learn from the inside of a training crate. Even an open top basket can be put to use. Or maybe a showpen provided it is secured at the base and tied in place. The top of the pen must be made secure.

JUNE

Through the simple utilisation of such appliances you can help to further the education of your babies. By being allowed to "take in" their surroundings from the inside of a well designed crate, or better still a portable aviary that can be secured in a simple way for the security of the occupants. Given such opportunities it is simply amazing how much they learn. Safe inside too whilst you are away from the loft although I personally prefer to be on guard in case of a roving cat. You have to protect your valued babies from predators, especially a neighbour's cat, but remember too, your neighbours are entitled to their pets in the same way that you are, provided your loft has been erected in accordance with local regulations and by-laws. Pigeon fanciers have to be very diplomatic where neighbours are concerned. It pays dividends in the long-term.

When I am obliged to carry out the settling of a few babies to the outside of the loft and especially when they are above the age of 28 days old, you have to be very careful, or you can lose them quite easily. Far too many young birds are lost simply because due care is lacking when they are first put outside. Again if you purchase a group of squeakers they only have to vary in age by the margin of between three to five days, and without using plain common sense you could be in trouble when you let them out for the very first time. It is surprising how strong youngsters can be within a matter of a few days, hence the need to take the simple precaution of preparing for this eventuality by encouraging your babies to develop "spot location", in a safe and simple manner. As well as preventing them from flying away from the loft before they have developed local knowledge. Personally I do all I can to prevent them from flying up and away when let out (given their freedom) for the very first time. Furthermore I continue to do so for as long as possible!

By following such a policy it helps considerably to prevent very young and immature pigeons from developing mass hysteria in flight. Many of the losses of young (baby) pigeons is largely owing to impatience. Today, more than ever before, it is so very simple to pick up a phone, and coupled with a credit card, purchase at a most reasonable price a group of a dozen squeakers from any of the well known commercial breeders. For those with ambitious plans for 500-mile racing, or even long distances you now have a wide choice. But at this time, from early June only you can with care as outlined hold most of your purchases. This way, you have your own flying-out stock. So that next season you will be breeding pigeons that fly out to your loft, which is a most happy situation. Patience, patience always — and next season's breeding programme will be much enhanced as a result. It is not imperative to race them. However, when your new purchases have become safely settled to their new environment, you can if you wish carry out a further-education programme by providing your new stock with the opportunity to negotiate several training tosses, say up to a maximum of 30 miles, or thereabouts. Lack of race experience will not scientifically or biologically prevent them from breeding sound flying stock for next year's young bird races.

The object of the exercise is to develop flying-out breeders as opposed to the ownership of non-flying stock that will have to remain full-time prisoners. Such a policy is two-fold. You will have the satisfaction of being able to produce, stock that can be raced in the young bird races. At the same time you will also have the opportunity of discovering the value of the parents. Furthermore, you will also have gained a great deal of experience and knowledge that will help you to improve your own ability as a breeder, and your own more ultimate knowledge as a stockman in a practical manner.

Many contend that young bird racing is a simple affair compared to the complications that such fanciers also insist is associated with old bird racing. I do not agree with this for one moment. On the face of it this may appear to be the case because in the matter of young bird racing you do not have the problems associated with old bird racing. Both the main systems employed be it Natural (flying hens as well as cocks) with mated pairs, or cocks only (full Widowhood) provide problems. The problems associated with Natural flying create confusing anxieties for the experienced fancier. Imagine the worries of the novice or sheer newcomer. However, I do believe that in the beginning, the Natural method is the ideal way to begin. Doubtless as you progress you will eventually be encouraged in the belief that it will be necessary to practise the Widowhood system of racing. It will most certainly overcome the problems of time and money respectively incurred to considerable lengths and costs that can be incurred with the Natural method of racing.

Training is imperative if you fly your pigeons on the Natural system. The advantage being that you have the opportunity of racing your hens. Nonetheless, the racing of young birds requires considerable skills. You may not have the complications of nesting problems, but young birds to be successful require a great deal of daily attention, and carefully planned training and daily exercise routine, in order to get the best out of them. Remember, unlike either the Natural system of racing, or the Widowhood system, wherein the sexual attraction is of enormous influence, the majority of young bird races are won to a perch, the love of home, and the daily feed, grit, minerals, and clean cool water. So in my opinion young bird racing requires considerable skill and dedication, as well as a great love on the part of the owner for his charges.

In Belgium, as well as Western and Eastern Europe, young bird racing has many professional-like exponents. Great sums of money are at stake. Young bird racing is for many most challenging. There are even those particularly in Belgium who only ever race young birds. How about that! Very large sums are frequently won by those with the greatest skill. It is progressing if slowly but definitely this way in the UK. Better bred pigeons the result of large investments by the larger studs have made this possible. So my advice to the newcomer is to do your utmost to get every youngster you purchase settled safely and surely to their new home. For those youngsters they have bred apply the same caution, in order to get them to the first race. Far too many young birds are frittered or wasted because of lack of patience in educating your young birds to be fully endowed with spot location of their home (loft).

A young novice fancier recently asked me how he could train one of his best bred cocks to defend his nest box and its mate from the intrusion of another cock. Since the bird in question was rated so highly by our novice friend it was a delicate situation. It was necessary for me to think quickly and at the same time try as much as possible to avoid upsetting my young friend. So to this end I recalled my own observations and experience over a long period of time. At the same time his friend, also a budding fancier, young in years, wanted to know what he should do about one of his very best bred hens that falls to pieces after taking a long time to lay its eggs.

My own experience of cocks that allow themselves to be turned out of their nest box, their home, without offering any resistance whatsoever is not good. They will not prove themselves good racers and therefore not suitable subjects to breed from no matter how well bred themselves. Such pigeons lack courage. Any racing pigeon without the quality of courage in abundance, is not worth his salt and should be discarded. Courage is one of the greatest assets needed to enable a racing pigeon to overcome the many obstacles it will meet over long distances.

With regard to hens that find the task of egg laying a great strain it is a sign of a weak constitution. As with any male that refuses to protect its spouse and home, hens that continously appear to mope or are very slow layers are weakly constituted and again should not under any circumstances be bred from. The exception may have been observed by some but these in my opinion must be rare indeed.

You will find that in both instances such pigeons will not stand up to the strain that long distance racing imposes. When they are unpaired they always look the part, yet immediately they are subjected to the strain of flying long distances they will be found wanting. Like those pigeons which cannot stand the strain of rearing a pair of squeakers, you cannot ever tell from their looks after a good moult which birds these will be. Any pigeon that cannot stand the strain of maternal duties will not stand up to the toughness of the work which it will have to face on the long, and often lonesome road home from four, five and maybe even 600 miles. One must forever be on the lookout for signs of weakness in the make-up of those intended for long distance racing. Remember what I have often said before, it is all a matter of survival of the fittest. Constitutional weakness and lack of courage render such subjects useless as racers, and thereby not worth keeping let alone used for breeding. Survival of the fittest is essential if you want to survive as a breeder of outstanding racing pigeons.

With the 500-mile races taking place within the next few weeks it is time to remind fanciers who may be sending for the very first time that often these races take ten or more hours to negotiate. The pigeons you select must be the best possible if you aspire to win a prize. Sending pigeons in the vain hope they will get home is rather foolish. Send your best pigeons and send them in the best possible mental and physical condition.

You must be prepared to deal with a somewhat tired bird when it gets home. Even an exhausted bird. Today there are many remedies and food supplements available. Pigeons that have to endure ten or more hours of continuous flight are in need of a pick-me-up and I cannot think of a better one than to beat up a whole new laid egg to which you add three teaspoonfuls of glucose, four egg cups full of boiling milk and two teaspoonfuls of brandy. Mix well until the glucose is fully dissolved, and while it is still warm give the bird an egg cup full at hourly intervals for at least two to three hours. It really works wonders. While you are at it, have one yourself! You can if you wish substitute sugar instead of glucose. Castor sugar is best, it dissolves smoothly and more speedily. At least that is my own experience.

Pellets too instead of hard corn are also very useful for the feeding of tired racers. But you do have to educate pigeons to eat pellets. Looking after tired long distance racing pigeons is most important if you aspire to the fascinations and rewards that distance racing brings. But remember too that distance kills so fitness is of the utmost importance, and so too is breeding ancestry.

JULY

Outstanding Beeders — Marathons Selection — Racing Ability — Peppiatt's 'White Hope' — Young Birds and Novices Heatwaves

With only a few exceptions, such as a few specialist cross-Channel races at the longer distances, for the vast majority the old bird season is over by mid-July. Without doubt many races on all routes have provided many difficult races this year, with good pigeons, that is to say proven distance pigeons, missing at nightfall on the day of liberation. Happily I learn through the regular grapevine that many missing favourites are beginning to find their homes. Doubtless others still missing will turn up eventually. Yet according to reports filtering through about late returnees these include pigeons that had previously flown or even scored in these same events.

Not for these enduring types another home! As we have witnessed over the years pigeon racing, and this is especially true of long distance pigeon racing, is without doubt a matter of the survival of the strongest and fittest and indeed the equally determined. Long distance pigeon racing is without doubt the most difficult aspect of our fascinating hobby. Only the strongly constituted and best bred pigeons according to informative ancestry will survive the extreme distance races. It also stands to reason that when the time comes to retire such stalwarts from racing these are the ones to look to for the continuity of the family by inbreeding or line breeding to the breeders within the family. The constitutionally sound will reproduce their genes and often continue to do so up to a great age. Physical strength, the legacy of a sound selective ancestry built up over the years, will always pay in the long-term.

A rigid selection based entirely upon racing and reproductive ability will find out the physically as well as the mentally superior. For those who have based their foundation selections upon good local talent, and had the good fortune to be able to purchase stock from the best racers and breeders in the seller's loft, they are assured of a good future provided they continue to select according to results. Each succeeding generation has to be culled no matter how gloriously they are bred. It is therefore logical to assume that to a large degree if such a policy is maintained without fear or favour you will probable cull less and less. Yet paradoxically according to a number of prominent Belgian fanciers it is necessary to purchase new pigeons annually. With this I do not entirely agree. Indeed nor do I advise, it is simply because I do not believe it wise, or necessary.

In support of my contention there are according to the many articles on Belgian as well as other Continental fanciers still a number who practise a considerable amount of line breeding and even intense inbreeding. But almost without exception these particular fanciers race their pigeons every year. It would probably prove much more difficult to continue a policy of inbreeding in the commercial breeding studs who do not practise any pigeon racing whatsoever. They justify their sales by continuously purchasing outstanding performers. For those who are able to afford expensive purchases from the top priced commercial stud's highly priced purchases of prolific card winners, my advice would be to concentrate your efforts upon a family, or strain, and practise your own form of line breeding-cum-inbreeding according to your own vigorous policy of progeny testing.

Forget what your initial stock cost you in terms of your pocket. Forget what they are bred from. Forget about all those remarkable wins of their predecessors, you will still have to work for the survival of the strongest pigeons. Purchasing the children and grandchildren of multi-winning short to middle distance prize-winners will not produce long distance winners. These are the most difficult pigeons to obtain. Whatever you purchase whether privately, or from the highly organised breeding studs, despite the ancestry as reflected in the breeding details (pedigrees) you still have to be fully prepared to continue to test out the progeny you produce even from the most expensive purchases. It is the mysterious lack of breeders of long distance racers that makes the task of producing long distance racers so difficult. Good consistent 500-milers are still in a class of their own. As indeed are the marathon type racers. If fanciers would make up their minds to test severely all the pigeons that survive to reach the race programme and not worry at all about losses because of their pedigrees no matter how well bred they may be, or how much you paid out, the sooner you will know how wisely you have spent your money. More or less this is the continuing policy of the survival of the fittest.

Recently I have been able to note in my own loft the various constitutions of old pigeons, that up to three to five years ago had given a good account of themselves in the races. Included were several which had five years ago and even longer proved themselves at the 500-mile distance. Additionally were a number which have reached the age of 10, 11 and even 12 years old. Despite the fact these very old ones have not been mated for these last three years many I am glad to write are still fully capable of reproduction whilst in certain cases others much less aged, appear at present incapable. This is yet another example of the strongly constituted as opposed to those weakly constituted. Whether it be for breeding, or for racing, the physical strength of both parents is of paramount importance.

Outstanding breeders are for this very reason of inestimable value. They are rare. Also they are worth their weight in gold, so much so that when a pair of outstanding breeders are sold from a very small loft that has been obliged thereby to breed sparingly, that loft can easily fall into decline if the sons and daughters are sold too, because of the extremely high value placed on the "golden stock pair". It is so very easy to be tempted to sell, but all too often a grave mistake. As many have found out all too late! This I emphasise to illustrate how important it is to conserve the best sons and daughters of the breeders and thereby enable you to line breed, not only to the racers, but line breed to the breeding lines it is within the family. Strange as it may seem not always easy to practise this unless you continue to progeny test, as well as having been fortunate to discover a positive breeding line (or pair of breeders) within your very own family. Lucky indeed are those who do. To do so is probably as difficult as winning a million pounds on the football pools.

Marathon-type racers
Whatever one's opinion concerning these extreme long distance races may be, one cannot help but feel great pride and admiration for those gallant arrivals in England from faraway Rome in the 1986 race. For my part I was, and in fact still am sceptical of the wisdom of promoting such events. Nonetheless I was thrilled to learn of the arrivals. It is a great pity the race was not won in race time. Nevertheless I made the effort to trace Mr & Mrs H J Kennett of Orpington, Kent, not only to congratulate them in their extreme confidence, but also for recognition of ancestry. It appears that the base of the family is the Bill Button pigeons whose pigeons I sold for the one time highly successful Ipswich fancier a few weeks prior to Mr Button's departure to Australia where he now lives. The late Glyn Parry presented Harry Kennett with six of his Bill Button-based family for training his young birds. And as is now reported has received all of the late Mr Parry's small Button of Ipswich family since Mr Parry's untimely death. There

is not the slightest doubt that Queen of Rome has a creditable background of long distance ancestry. In due time I hope to be able to produce, to the full extent of the information obtained, her total ancestry. At least to date there is one more bird home than was recorded in the 1910 race from Rome, in which two homed, and the time taken by Queen of Rome has been reduced from 31 days to 11 days 3 hours 42 minutes to be precise, no mean performance by any standard. She was bred in 1981 and has always been a great favourite with the Kennetts.

Selection
In selecting pigeons whether your inclinations are towards short distance races, middle distance, long distance, or marathon-type racing you must recognise that the most important aspect of selection is based upon racing ability. There is not the slightest doubt that here in the UK there are thousands more pigeons that appear to be sprinters. It is more than possible that it may increase in its popularity. Doubtless this is owing to the great surge towards Widowhood racing. Mostly the success of Widowerhood (to give it its correct title) is largely owing to the system, and where exceptional success is shown owing to the skill of the manager or owner as the case may be. In this country it is far easier to win a large number of 1st prizes by an exceptional racer than it is in Belgium where the numerical strength of the Belgian club often runs into hundreds rather than in dozens (or even less) as it does in the UK. Although there is an increasing number of Belgian clubs that are becoming less numerical membership-wise owing to the trend for economic reasons toward the need to build blocks of flats rather than expensive houses, which fewer would-be fanciers can afford to buy, or even rent. Nonetheless the normal club membership makes it far, far more difficult for a single pigeon to amass ten, 15 or even 20 or more 1st prizes as can be achieved in this country. Yet whatever your interest, be it sprint (short distance), middle distance, long distance, or marathon-type racing (600 miles to 1,000 miles) you must learn to recognise racing ability.

Racing Ability
Although it is almost impossible to define the factors relating to racing ability, we will not go far wrong if we base our judgement upon racing performance.

Conversely the quality of a breeder will be assessed upon its ability to produce pigeons that are successful as racers. Progeny testing is the best and safest way to prove one's stock.

It is most certainly the more reliable method of assessing your breeders. In plainer terms as I have so often mentioned before "Old

John Basket" is the best way to progeny test. The successful fancier (breeder/racer) has to possess great patience, acute powers of observations, a meticulous love for recording details (even despite being endowed with a good memory), and a most observant eye which not all fanciers seem to possess! Experience counts for a great deal but this only comes with practice, and years too, which alas pass all too quickly. In the successful cultivation of the racing pigeon you have without doubt to rely upon the basket in making your selections.

Only the fittest and strongest will survive. Sounds simple enough but unfortunately it is often not enough only because fanciers will place more faith in a wonderful pedigree without learning to recognise the need for vitality in your subject. Good health as I have witnessed in many pigeons, is not the same as vitality. In the exceptional pigeon you have both.

Now that old bird racing is practically over, we have to accept that the weather in recent years over the British Isles and indeed the parts of the Continent that our pigeons raced from has been like the curate's egg proved good in parts! Nevertheless outstanding pigeons emerge no matter how difficult the races prove to be. These individual types that show by their consistency and outstanding ability to race home despite the variable weather en route are the ones we all seek. These are the pigeons to help you produce a good reliable family. Noticeable at this time in particular has been the number of outstanding racing hens that have emerged, especially in the greater distance events. The outstanding racing individual that repeats a good performance is hard to come by, especially at the distance. Gladly too it now appears that there is a greater interest in the longer distance racing events. In this the great National clubs of this country can claim a great deal of the credit as can the long distance specialist clubs. The prowess of their successful members with outstanding performances produced by outstanding individual pigeons, be they cocks or hens, raced Natural or Widowhood, inspire others, thus helping to stimulate interest and a far greater intensification of the desire to produce pigeons capable of racing the distance.

It is pleasing also (at least for me) to note that the desire to obtain pigeons that are capable of producing pigeons that will race long distances is seemingly far greater than the actual supply! Pleasing to me because it shows the present trend towards long distance racing.

There have over the years been a number of outstanding hens that have proved supreme, and especially those which seem to excel themselves from a particular race point. Fortunately even in this present age, with the racing of cocks only being the rage, outstanding hens continue to emerge. Doubtless they will decline in numbers overall as more and more fanciers take up full Widowhood in this country for the sprint to middle distance races. Happily there is growing interest in greater distance racing and in view of the successes of hens in these events so far as long distance racing is concerned the ladies are likely to continue to emerge from time to time.

Recently as some may have noted the editor more or less "invited" me to try and deal with an outstanding hen of over 60 years ago. Excusably in his reference to this particular pigeon he referred to her as a pure white which she was not. You will obtain the gist of it all if you refer back to The Racing Pigeon under "Opinion" headed "Improbable or Impossible?", wherein it is stated, "such margins are not impossible". In all fairness the races in question were vastly different. Whereas the 1922 Banff race was practically a disaster the recent Stonehaven LNRC race was a comparatively easy race and with several thousand birds competing! There is no real comparison between these two events and different velocities between the two first arrivals and the third and fourth arrivals in each event which as it happened were both London NRC events, and both representing the second of the three Combine OB Classic races proper. The one that White Hope won was flown from Banff approximately 425 miles, the 1987 version being flown from Stonehaven, approximately 375 miles with upwards of 7,000 entries. Winning a bad race as opposed to a faster easy type race by a margin of four hours for the Banff race is a different kind of result than was the difference between the first two pigeons and the third and fourth arrivals from Stonehaven.

The pigeon referred to was Charles H Peppiatt's White Hope which was, in fact, a speckled-faced gay chequer pied hen ringed 17.09. She was raced by Charles Peppiatt, whose loft was on Haverstock Hill, Chalk Farm, London NW, and who flew in the then very old established but now defunct County of Middlesex HS, also London Premier, Islington HS and Kentish Town DHS, as well as East London Fed. As a young bird she flew York. As a yearling she flew and won what was called the London Premier 1,000 Classics from Darlington (214 miles). She also won 2nd London Premier Derby from Perth (360 miles) as well as equal 2nd in the Perth Open race. In 1919, 09 as she was then known, won 1st East London Fed Banff, 1st London Premier £5 Championship Club, 1st County of Middlesex, 1st Islington FC, 2nd London Premier FC and 2nd Open London NR Combine Banff winning £125 in pools and prizes, quite a bit of money in those days! In 1920, still known as 09, she won 1st in all clubs and 3rd East London Fed with 2,167 birds competing but I cannot trace the actual race point. From Banff she won 1st County of Middlesex, 1st Islington FC, 1st London Premier FC and again 2nd Open London NR Combine Banff winning about £115 pools and prizes. In 1921 by which time she had been christened White Hope, for as Charlie was wont to say "hoping to win the Combine this year — third time lucky with my

White Hope" and the name stuck.

I met Charles Peppiatt several times and in later years got to know him well. Charlie's cousin Daisy married a Mr Cox and they became the parents of Iris Cox who married my wife's eldest brother. Later when I was secretary of London Social Circle before the last war, Charles Peppiatt regularly supported my London Social Circle activities, especially the river trips down dear old Father Thames embarking at Windsor and then down to Marlow and even beyond!

Mrs Daisy Cox had a brother-in-law who kept and raced pigeons at Edgware and with whom I exchanged pigeons way back in the thirties. Some of his pigeons were descendants of Charles Peppiatt's family that were based upon the early Logans, Barkers, Vernal Hansennes, Wegges and a noted distance winning family known as Boddimead's Lerwick strain. It is doubtful if anyone can recall this strain today. Other bloodlines used by Peppiatt included the original Tofts, the Van Cutsems (through Mr Cuthbertson of Wanstead) and Dr Geo P Alderson's Brain Pan family, a noted and highly successful family in its day. Brain Pan was a pigeon that was for various reasons almost eliminated from the loft at Turton, near Bolton, Lancs. An unknown at that time and criticised by Dr Alderson for several faults, but Simpson, the doctor's loftman, observed it is not always shape and good looks that count, it is what's in the brain that counts most! It is easy to see how the pigeon got its name, for as a breeder it proved phenomenally successfully. Every nest almost produced a pan full of brains, so much so that the doctor was eventually forced into describing his lofts as the Brain Pan Lofts, and in due time it housed the Brain Pan strain! Simpson's discreet remarks saved a good pigeon. Among his many excellent children was a hen named Shy Lass that became a legend in her earlier years. As a breeder she proved extraordinarily successful. However, I am drifting away from White Hope which by 1921 had won 2nd Open London NR Combine three years in succession.

For her 1921 racing successes White Hope was awarded the National Homing Union Gold Medal for the most Meritorious Performance of the Year. She also won the Lloyd's News £5 Championship prize. In 1922 White Hope finally enabled Charles Peppiatt to realise his greatest ambition, winning 1st Open LNRC from Banff. Furthermore she won it by a margin of four hours. There were 1,042 birds competing and she won about a £100. For this great performance White Hope was awarded the National War Pigeon Trophy and Gold Medal, described in The Racing Pigeon, "for the finest performance ever known in the British Isles". So far, after many hours I have still not discovered the history of this trophy! Again White Hope won the Lloyd's News £5 Championship prize. In 1922 White Hope was retired after flying Banff again and winning pools to £5 in Combine and other pools to £40. Her huge oil painting was commissioned and hung behind Charles Peppiatt's desk chair for as long as Charles lived. I have often wondered what happened to that painting as well as how that National War Pigeon Trophy came about. Actually it could have been a trophy presented to the Fancy by the War Office to commemorate the work of the British racing pigeons during the 1914-18 war, which was commanded by Lt-Col A H Osman.

In concluding the hard work of researching and compiling as much as I can the history of Charles Peppiatt's White Hope it reminded me of the time I was asked to sell a smallish loft of pigeons by auction. When the details were received it amused me to note that the children of White Hope had been extensively used by the person who had supplied my client with his foundation stock. Both the pigeon White Hope and her breeder had long been dead. You see I knew that the three times 2nd Open LNRC Banff winner who crowned her career with 1st Open LNRC Banff in 1922 was a barren hen. One of the pigeons that beat White Hope was bred and raced by my old friend Sid Magee who lived at Ponders End, Enfield, and whose daughter worked with me at Hallmark House for 15 years. There are still a number of fanciers who will recall when having clocks set for either NFC or NRCC at Sid's home seeing that 1920 Banff Combine winner in a glass case as a result of the skilful work of a local taxidermist. Sid Magee was offered the sum of £10 for his winner but steadfastly refused to sell his champion. Unfortunately she was shot down by persons unknown with a high power air rifle soon after he had been offered that princely sum. One thing is certain, and that is this, White Hope may not have produced any children, but she most certainly establish a LNRC record that will be hard to equal.

Great pigeons are always hard to come by. The harder you are regarding selection the more likely you will be in producing pigeons of greatness. The adjectives "great" and "greatness" are of themselves, a permanancy of description. In this I use these adjectives in illustrating such qualities in the establishment of racing pigeons of long distance racing ability. For my own part I am only interested in long distance races. Unfortunately these past ten years have proved difficult for me to be able to participate. My passion for the production of the long distance racing pigeon is as intense as ever. However, the ability to prove the breeding is mainly restricted to others to prove out the wisdom or otherwise of those that I have been able to produce.

Today, however, there are it appears many more who prefer to specialise in short distance, 50- to 120-mile racing. Equally, there are many who have a great desire to succeed in races between 130 to 230 miles. To succeed in either or both groups requires much skill and dedication.

The group that I personally am intensely dedicated to, is the group of pigeons that are capable of racing from distances of 300, 400, 500 and 600 miles. The only successes I can refer to during the last ten years is mainly success achieved by others with the descendants of those I have bred. There was a time when the "well bred" pigeon simply had to be bred from pigeons whose ancestry often included pigeons that had either flown successfully, or were either the children or grandchildren of top class racing pigeons that had proved themselves in races of 300, 400, 500 and/or 600 miles.

Years ago the famous founder of the Osman strain, and others too who also specialised in long distance racing and founded a strain that bore their names, ie N Barker, Logan, Toft, Spangles (Oliver Dix), Savage (of Stanmore), Bruton (of Palmers Green), Fuller-Isaacson (of Muswell Hill) and others with similar aspirations, produced annually pigeons that excelled in the longest type of races.

In those seemingly far off days it was therefore much easier for fanciers whose aspirations were to produce pigeons of long distance racing ability and heed the advice of the above fancier, and A H Osman, in particular, who was always stating in his brilliant articles the importance of purchasing their foundation stock from a winning loft. In this day and age you have specialist fliers who excel in short distance races only. There are a greater number too who excel in races from 130 to 250 miles. For those who excel in the races flown from 300, 400 and 500 miles the number who are successful is greatly reduced. If you include those who succeed in all four distances, 300, 400, 500 and 600 miles, their numbers are indeed noticeably even smaller.

For those prepared to heed the advice of the highly successful famous fanciers I have mentioned, it is as well to qualify the statement, "buy from a winning loft", by making certain you purchase foundation stock according to your aims and ambitions.It would prove both wasteful of effort as well as expensive if you fail to realise that the commercial studs cultivate strains, and for families that will succeed at the shorter distances, especially if you fly Widowhood or other forms of Widowhood systems. Happily, today there does appear to be a greater interest in the longer races.

For those who specialise, the rewards are relatively enormous compared to the earlier years of the sport, and even the 20 years post war! It is for this reason strict rules must be enforced and upheld in a highly efficient and vigilant manner. Regarding those fanciers whose ambitions and aspirations to succeed in greater distance racing they are like their pigeons, an elite group.

Doubtless there is a great deal of skill to succeed in the shorter distance events. Equally too, it is today pigeons for courses, the sprinter type of pigeon. It also requires a great deal of skill to succeed consistently in the races from 130 to 250 miles especially at Fed and

MONTH BY MONTH — in the loft

Open race level. The sport of pigeon racing today has changed enormously. The commercial studs who cater for all interests have been largely responsible for these changes by imports. You can today obtain strains that will meet the needs of most fanciers' ambitions. Fashion, too, dictates. Nevertheless certain strains are more popular than others. Who would have ever thought that the Busschaerts would have dominated for the past 30 years in the way they have? Doubtless the advent of Widowhood racing and the various methods of Widowhood-type racing has helped enormously in successfully establishing the groups of short distance, and middle distance fanciers. Yet even the Busschaerts and the Janssens, despite the enormous success in short to middle distance races, these two strains have achieved, they fail frequently, when tried at the longer distance races. There are exceptions we know. Both strains have excelled when crossed with established long distance strains. For example, Delbar, Cattrysse and Stichelbaut and a variety of Continental strains.

There is no doubt whatsoever, that today more than ever before, the hobby of pigeon racing is becoming more professional than it ever was. Nonetheless, the long distance champion racers are as difficult to come by as they have always been. Distance kills. Stamina and constitution must never be in doubt. The most important pigeons are those who can justifiably be described as "breeders". These rare types are prepotent. When discovered guard them with your life, waste not a single egg. Prepotency is all powerful and fully responsible for the production of strains that produce outstanding long distance racers continuously as history reveals. Prepotent pigeons are "worth their weight in gold".

The famous J W Logan discovered the value of "line breeders". Several of his best distance racers (his sole aim was long distance racing pigeons) were the result of line breeding. A glorious example was J W Logan's production through line breeding of Champion 1826, the marvellous red chequer hen.

This indeed was a glorious example of prepotent line breeding. The "breeders" within a family are "prepotent". Line breeding is the quickest way to exploit prepotency. Logan's selected male was 963 and the production of 1826 the famous San Sebastian King's Cup winner, was bred actually as follows: Logan 963 was mated to Logan 673 and produced Cock 1728; Logan 963 was mated to Logan 8260 and produced Hen 1315; and 1728 and 1315 were the parents of 1826 who was the 8th prize-winner from San Sebastian, Spain in 1921 and won 1st prize San Sebastian and King's Cup in 1922. To be described as prepotent is to be very powerful for breeding in a most superior manner.

Another of Mr Logan's prepotent males was the famous deep red chequer cock known by all Logan enthusiasts as Champion 69. Bred

James Bruton (left) outside his lofts at The Limes, Hedge Lane, Palmers Green circa 1930.

in 1919, and as a youngster was only trained 50 miles, as indeed was the famous hen 1826. Champion 69 flew the San Sebastian National as a two year old in 1921 winning 19th prize. In the extended pedigree of No 69 Mr Logan detailed 161 different individual pigeons that had passed through his lofts, tracing back 50 years. Of those 161 birds, 28 were the result of direct importations mostly through his famous English born agent, Yorkshireman Northrop Barker. One hundred and one were bred at East Langton from directly imported Belgian birds. A dozen were obtained from English fanciers. Another dozen were bred at his home at East Langton from the birds obtained from English fanciers but crossed with birds bred at East Langton. In the genealogical tree of No 69 it revealed that he went back in 92 branches (or lines) to Logan's Old 86, whose photograph thankfully is still displayed on the front page of each issue of The Racing Pigeon, where it has appeared on each and every issue, since the No 1 copy published Wednesday 20 April 1898, price one penny! Old 86 was bred in 1879 from B6 (N Barker's Montauban) 9th prize-winner Belgium National of 1878. The dam of Old 86 was 646, a Georges Gits hen sent by Mons Gits to J O Allen, who gave her to Mr Logan. Old 86 won 1st prize in his last four Continental races, winning twice 1st prize Rennes, 310 miles in 1884 and 1886 and twice 1st prize La Rochelle 444 miles also in 1884 and 1886.

There were also 68 branches to Old 86 when he was mated to 163, a blue chequer hen bred in 1881 from D2 and famous N Barker bred hen B16, 163 flew Marennes 463 miles twice. The lines to D2 and B16 arise 196 times. There were 88 branch lines to blue chequer No 22 that was bred by a friend in Antwerp in the year 1871, and sent to Mr Logan when he lived at High Hazles, Sheffield. No 22 was a squeaker at the time of his dispatch from Antwerp. In 1873, No 22 flew from Ostend to Sheffield twice. D2 mentioned above was a son of D1 that Mr Logan purchased from Mons Debue, a stonemason of Uccle; the red chequer (Debue 1) took a good prize in a very bad race of the Belgian National from Auch, in 1879. Mr Logan at that particular time journeyed to Brussels to see Northrop Barker's birds arrive from the race, the weather was atrocious, many birds lost. Debue's red cock won 31st Open National. Mr Logan was so impressed by the performance in view of the weather he asked N Barker to purchase.

The negotiations took five or six weeks. No doubt the delay was engineered in order that Mons Debue, could obtain eggs, possibly with more than one hen. Strangely D1 never really impressed Mr Logan, but at the time of a subsequent visit to Barker's loft, Mr Logan noticed a very nice one-year-old red cock that bore a striking resemblance to Debue's prize-winning Auch pigeon (D1) that N Barker purchased. When told it was a son of the Debue Cock Mr Logan immediately replied: "You are not a bad judge, my friend; I like that bird very much indeed and would rather have him than his father". Subsequently a bargain was struck, and D2 as he was immediately called (since his sire was listed as D1 although never used) was mated to probably the finest hen Mr Logan purchased from Northrop Barker known as B16, being own sister to N Barker's Montauban, the sire of Old 86. When B16 was paired to D2, they were responsible for blue chequer hen 163, also red chequer hen 187, both flew La Rochelle 444 miles in the early 1880s. D2 and B16 also bred a wonderful red chequer cock, the well known 105 that Mr Logan retained for stock until 105 died. This is a typical example of prepotency so important if one aspires to produce his own family or strain.

There always is a great deal of interest shown by fanciers on how to obtain the very best possible condition in a racing pigeon. In particular never more so than at the commencement of young bird racing when the newcomers to the wonderful hobby of pigeon racing find it upon them for the very first time. Unfortunately a great many fanciers imagine condition is brought about by some potion, or secret elixir. Good health is the real secret to success and this natural phenomenon is not brought about through the cultivation of pigeons that are weakly constituted. Medicines will not transform a weakling into an athlete, but that is what a first class racing pigeon really is! Constitution is of the greatest importance in the breeding of the racing pigeon, especially the long distance racer.

There is now a definite swing to longer distance racing. Even the greater distance fanciers are now fully catered for with the advent of marathon-type races. British fanciers can compete against the Continental fanciers. Personally I do not see why British fanciers cannot achieve enormous success in International racing.Most certainly success will be achieved if British fanciers cultivate the right type of pigeon. A sound constitution is of the greatest importance. Self-reliance is another important factor in the make-up of the long distance racing pigeon. Constitution of the body as regards health, physical strength and mental character are for the purpose of this advocacy, developed successfully through a sound environment, as well as cultivation of the genes for distance racing is intensified. Qualities of physical and mental strength really have to be developed and exploited to the full. Outstanding racehorses have great constitutions and strength of character. Wonderful creatures, the likes of Dancing Brave, Reference Point, Mtoto, have these rare qualities, and their respective trainers who have made a lifelong study of their equine charges have trained and managed their charges to the fullest extent through their great knowledge in order to create such legends. Such knowledge is based upon observation and experience, more or less in the same way that Jim Biss has nurtured, managed, trained and brought out the very best with regards to his great ambitions for long distance pigeon racing in the United Kingdom.

From the telephone calls and conversations with callers it is quite evident that YB racing is for many a great attraction. For the newcomer, those without any previous experience whatsoever, it is a great and exhilarating feeling, and they simply cannot wait to get started. At the gathering of members at my local club, on the final date for the YB nominations, old-timers, novices and those young in years, consequently with lesser experience were obviously all very excited at the prospects of yet another YB racing season. Several it seems had suffered a few losses after flights around the loft. Others through a bad training toss, but overall the majority had a goodly number ready for racing according to the total number tallied. Owing to the extreme heat, clear skies (a sky that is without a vestige of cloud, and the sun in full glare), as well as easterly winds, these conditions can provide headaches for fanciers and prove a severe handicap to inexperienced YBs. Broken cloud provides cover and guidance for youngsters. However, from what I could gather from the varying conversational exchanges it was quite obvious that those who had given their youngsters a very big initial jump for their first toss had been extremely lucky.

For safety's sake and especially for the raw novice, I always advocate short easy stages. It so happens the weather generally in these parts has proved tricky for those fanciers who have not had too much experience, and are quite unable to evaluate skies and cloud formation also those who overlook the dangers that easterly winds can provide, they have come off far worse. In and around North London area several lofts have been completely emptied of their first and second rounds. In many instances these training tosses have taken place too late in the day. Evening tosses are alright when the weather is ideal, and only the old and experienced fanciers know all about the dangers of late training tosses during a heatwave! Young bird training is vastly different to OB training. It is much more hazardous, as so much depends upon the weather. With the OBs you can train early in the day and not have the fear of losses that heat and clear skies can bring about with a kit of inexperienced youngsters, especially in easterly winds. Even inexperienced OBs can be upset greatly during heatwave conditions.

There are so many different ideas put forward by fanciers. One fancier had jumped his babies 35 miles first toss and got them altogether in very good time, so he took them 45 miles next time, and experienced a very bad toss with heavy losses. On checking back as I did, I found the weather that particular day was not at all good, extremely hot, a sky completely clear of broken cloud with intense sun glare and as well as an easterly wind. Not all YBs that are lost this way are duffers, not at all. Years ago you had the facilities of the railways for training pigeons. Those of us who were able to use the railways know only too well what a marvellous service this was for both fanciers as well as our pigeons, especially young pigeons. Today in place of the railways we have the motor car and various services for road transportation of our race teams for privately organised training programmes, as well as club and even a few Fed training schemes. But these latter type services invariably provide 25- to 35-mile flights only. Therefore, my suggestions for short training tosses can easily be carried out when your youngsters are nine and ten weeks old. A dozen short tosses is my firm advice for novices up to 20 miles. After this has been carried out you can then complete your training with say a 30-, or 40- and maybe, if time permits, a 50-mile toss before they are 12 weeks old. But always pay great attention to the weather forecasts and the wind direction. Avoid like the plague easterly winds when training. Training as I wrote previously, with a dozen tosses up to 20 miles, can be quite adequate if Feds commence their first races around the 50-mile mark.

Unfortunately many Feds commence at much longer distances, some as far away as 100 miles and even more. This is too far for babies, although I know pigeons as young as eight weeks old are frequently reported 100 miles away. The best advice I can give is train your YBs as early in the day as you possibly can. At the same time make certain they are ready for a feed when they reach home, and this way you will be more certain to trap them as soon as they arrive. Young birds learn quickly and do not forget easily. Consequently if you encourage them into undisciplined habits they will be hard to break from these. Young birds that are trained well and taught well on how to trap once they arrive from a race will bring much joy.

A good many fanciers are much less interested in YB racing due largely to their greater ambitions for success in long distance racing. Nevertheless I do not consider it is a bad thing to educate your YBs by racing them if your time and circumstances permit. Although history has shown that many good OB winners have been unraced as YBs, yet equally many more have been well schooled at the racing game as YBs and gone on to become champions. After many years at the game I consider that if you are not keen to race your YBs, or that family commitments make it difficult then it would pay you well to consider training your current year's stock, say up to 60 miles once or twice. It helps to develop their physique as well as their minds. But remember at all times good habits are of the greatest importance. Pigeons are no different to young children, both are creatures of habit. Therefore, it is of the utmost importance that you instill good habits at the earliest possible stage of their lives. Such habits acquired through a firm discipline when young will influence their minds enormously. Its effect, if meticulously carried out, will last a lifetime. In looking back over the years because I was greatly interested in long distance racing I tended like so many others with similar interests to overlook the education of the youngsters. My recollections and experiences lead me now firmly to the conclusion that I was wrong.

I know there are many who will and can pinpoint good examples of outstanding racers that were never raced at all as youngsters, but equally I am convinced from what I have read, and from the many who have quoted me examples of successful YBs that have gone on to become outstanding racers that education for YBs is the best possible way to build up a racing family. Today because of the greater interest in Widowhood racing only, then it is in my opinion imperative you race a team of YBs even if only to race the hens out to the longest YB races, stopping the cocks as soon as they reach the 100-mile stage. Better still keep several pairs of racers to fly on Natural system, and a separate loft for the widowed cocks. For those with small lofts and limited facilities the Natural the system still wants a lot of beating. You only have to read thoroughly feature articles about the winners of the big races to learn for yourself the value of good racing hens. Often too they will show themselves, given the chance to compete in YB races.

You do not have to flog your babies to death in order to prove them. This is not a good policy. Young bird racing is a matter of education for the future. Particularly important too is the need to note the most consistent. The triers are far less abundant than you can envisage, only a thorough training and a good routine will bring to the fore the really outstanding pigeons. The future of one's loft depends entirely upon a thorough search for the exceptional pigeon. The ace racer, the champion of champions, these exceptional racers are not all that easy to find. A steady programme of training and racing YBs, and a determined effort and constancy of thought on your part to avoid any hindrance to their natural advancement and physical development, must be the paramount aim of the beginner who has ambitions to produce a family of long distance performers. Provided you have the right kind of stock from the beginning then a sound and constant policy of patient understanding of the requirements of a sound physical development, coupled with an intelligent approach in the maintenance of their health and happiness within the loft that only a really considerate and understanding environment will provide, will in the long-term be fully justified through the results that are bound to follow. A sound environment implies the need to provide warmth and comfort and fresh air when it is hot. Freedom from lice and red mite which abound when the sun is at its height. Clean water, regular baths and a dry floor. Clean corn well served in clean hoppers. A variety of grit that is served in clean utensils and like the corn in requisites that are suitably covered. All this is part of a good environment. Clean nest bowls, clean nest boxes, with clean nesting material are equally important.

Based on my own personal experience, not all outstanding racers were, or are, successful racers as YBs. History has shown that many who were apparently backward as YBs eventually developed into top class performers later on in their careers. You have to take note of the quite unobstrusive yet consistently regular racer. Those that show great love of home and are always eager to trap when they arrive. These are the types to look for. As you gain more experience you will notice certain pigeons always look clean with close fitting silky-like feathering. Outstanding pigeons always look clean, and most definitely have a good temperament and a personality of their own.

Yet intelligence alone, or tameness of the individual alone, or its cleanliness in appearance alone will not indicate the outstanding racer of the future, but it will help a great deal to observe such qualities since it is my belief that such attributes can be found in the outstanding racer. And such outstanding pigeons, when physically fit, are the more likely to excel as a racer when they reach a maximum of super fitness coupled with a state of mind that is motivated by passion and sensation which is obviously to my mind that the late bred hen carrying four nest flights and feeding seven day old squabs won the recent Thurso 500-mile Classic of London North Road Combine in a heatwave. It was motivated by a passion, love of its young for the first time in its young life and must have been very well bred too. Thus another record for a Combine Classic 500-mile race has been estabished that may be very difficult to surpass. Unless a new avenue has been established for the entry of late breds that are only nine or ten months old is predicted as a result of Mr & Mrs Mick Harvey's temerity to enter this little hen, now lovingly named Rosie's Delight. As Ray Jones, Combine president said to me, "That's a record that will be hard to beat, as well as a sensational one for Combine racing."

A heatwave will play havoc with the young birds. Extreme heat creates problems for the inexperienced youngsters. Intense heat produces haze, and it often creates high cloudless skies and these are very dangerous indeed. Unfortunately we have no control over the weather, and the difficult conditions that accompany extreme bouts of heat. Even for experienced old birds, as some races at the distances have shown already year, the extreme heat creates problems.

Many losses of outstanding racers have been brought about as a result of extreme heat, and complete loss of cloud cover, coupled with a blazing sun scorching down. Imagine what it can bring about for the innocently inexperienced young bird. It is for this reason that the more experienced fancier always insists that the lesser experienced should refrain from "putting all his or her eggs in one basket". For those who have ambitions to excel in the long distance racing events then caution is of the greatest importance. Try and learn to evaluate the variations of cloud formation. To experience a disaster after one's efforts to produce a worthwhile team of young birds with future long distance racing in mind can be most discouraging. Precaution is of the greatest importance. Many good pigeons have been lost through the fancier's own lack of both patience and foresight. For the young fancier commencing to build up a stock un-warranted losses will spoil his or her chances for progress in the next two years. Mature, experienced yearlings and two year olds are of vital importance.

You will be greatly reduced numerically unless you take the greatest possible care to hold as many of your young birds as you can. For those who are ambitious to succeed in long distance racing, it is imperative that every possible precaution is taken to conserve numbers rather than deplete them.

Believing those that are left are the best is not always a certainty. Those fanciers with established lofts, and in many instances, with large teams to work with, can to a certain extent, afford to take chances. They can and often do race their youngsters in every race on the card, plus a few midweek races thrown in. But the efforts required can often lead to severe losses especially during periods of intense heat. Older hands with fully established lofts, and extremely large teams of young birds can of course take far greater chances by subjecting their young birds to an excessive number of races. Relying upon numbers, say 100 youngsters, in the hope that one can finish up the young bird season with a dozen young birds seems a high price to pay.

As I type these lines, the sun is scorching down, clouds high, and consequently little or no cover whatsoever. In these conditions young racing pigeons can quite easily run into trouble. Furthermore, under such conditions they will easily panic. Being creatures of habit, and fully accustomed to their daily provisions, water, corn and grit (waited on hand and foot, as they so richly deserve), they have no other thought but to get back to the loft, where they know they will be well catered for. However, the weather is the great decider.

One might almost state blazing skies, extreme heat, and a scorching sun shining fiercely, can make their homeward-bound journey a cause for great anxiety. So much so that they can become panic stricken. Fearful of the unknown, as young children are when they lose sight of their mother when they accompany her on weekly shopping expeditions. If fanciers, especially those with little experience, would stop and consider the possibilities that are stacked against the progress of young immature racing pigeons many of the awful losses which seemingly are all too often reported nowadays would be greatly reduced. It does mean of course that fanciers must take steps to heed the advice of those who know a lot more about the pitfalls that beset the uninitiated. In matters of training and exercise of young pigeons, it will also mean that a more serious study of the weather, wind direction (easterly winds — especially direct east, ESE, and SE when the sun is scorching). When such conditions arise it spells disaster for old bird racers let alone the young and very inexperienced. The power of the sun is of great importance. We also know how important the sun is in enabling the pigeon to get a line for its home. However, too much sun can prove very difficult for young pigeons. Especially is this aspect of the weather of paramount importance in record-breaking heatwave conditions.

Other features to learn or observe is cloud formation andBarometer readings are of great use to the clever pigeon fancier. It amazes me that more fanciers do not include a good barometer in their equipment. There is not the slightest doubt at all that the best convoyers are those who have taken the trouble to study cloud phenomena. Such knowledge coupled with an intelligent use of weather forecasts especially Weatherline and, where possible, sound and reliable line of flight information will help considerably in the success of a liberation. Convoyers too can learn a great deal from barometer readings. For the novice fancier and those with a record of unsuccessful liberations, note the following: "When the sky is bad, dark and ominously threatening, and the wind E or NE invariably visibility is certain to be poor".

Visibility at the point of release should be good. Extreme heat and a strong powerful sun without cloud cover is not good, broken cloud is a great advantage and therefore is of the utmost importance. Avoid even the shortest training tosses when the sky is completely filled in (overcast) and consequently creating bad visibility. Remember too the ancient old saying: "When the wind blows from the east, it is neither fit for man nor beast".

This period of the year is all important for the novice fancier as well as the lesser qualified. So I am trying to help as much as I am able the many who are worried about the problems of establishing a new loft of pigeons as well as those who have already embarked upon such a venture during recent times have been finding it extremely difficult to report any success in establishing even a small loft numerically.You simply have to take the trouble to learn about the weather and all its many implications. If you don't all your efforts, and the money laid out will appear a terrible waste. And it really does not have to be that way.

With heatwave conditions it is equally important that young pigeons can be educated to fend for themselves when crated, or basketed. Far too many fanciers are slapdash when it comes to taking the trouble to train young pigeons to drink when basketed. A crate is more usual now, even if you can afford the ever increasing price of a wicker basket, or have the patience to take your place on the basketmakers order book, for they too, have problems finding the right type of willow cane, not to overlook the great shortage of good craft basketmakers. Even so I cannot stress too much the importance of educating young pigeons to become fully accustomed to "living" in a crate, or a basket.

It's a bit late if your young bird programme has started but it is never too late if you have not considered this important aspect of a young racing pigeon's curriculum. Just a few nights spent in a basket or a crate, I prefer a crate in this day and age, followed by a few short periods during the day time will soon enable you to draw your own conclusions as to the benefits your pigeons will enjoy once they have gained the confidence to enable them to stand up to their natural anxiety and attitude towards close confinement. Handle them with great care. No overcrowding whatever. Plenty of clean wood chips on the floors of either type of container. Good clean straw, or even hay if clean and sound. Do not ever be persuaded to accept or use either of these two items for floor dressing unless they are clean and dry. They simply must be free of damp smells, or you can create further problems for yourself as well as your pigeons. Remember whatever you decide to use, do not ever use damp or obnoxious smelling straw.

AUGUST

YB Losses and Fitness — Late Breds Transporters — Early YB Training — Nest Boxes — Separation — Champion Major

Every year the young bird season is for many proving to be a very frustrating affair with enormous numbers missing. Many of course return in due time, but many more do not ever return. They are lost for evermore. The majority of fanciers do not seem interested in sending for those that have gone astray. Maybe it would be worth introducing a rule that any young bird that decided to take up residence elsewhere automatically belongs to the fancier into whose loft he or she decided to enter and to whom a home is offered. Each season there does appear to be a considerable increase in the numbers lost. Maybe we are going wrong with our breeding policies. Too many young birds are going astray. Unfortunately many fanciers are not willing to accept a gift of a pigeon in this fashion. Maybe the present mode of transportation is at fault. Or is it due perhaps to the lack of organised training facilities by the Feds coupled with a complete lack of co-operation between the Feds when racing routes and programmes are being considered?

The major losses seemingly occur in the first three or four young bird races. Personally I really do believe a great many of these losses would be reduced enormously if we had a succession of low mileage races instead of the trend for increasing the distances as quickly as possible. The initial races for young birds commence at a far too great a distance. Lack of training facilities also adds to the carnage. Too many young birds are turned into strays because their training is too sparse, even for some, none at all! Far too many young birds are turned into strays because they are not given sufficient exercise, and thus are lacking in flying power. In plainer terms they are without muscular development. Equally a goodly number are overfed and consequently are too fat — no muscle, all flab! Success in young bird racing is for many a very series matter so far as the desire to win races is concerned. Yet a great deal is lacking. Large numbers are kept in order to overcome the sheer inadequacies of the policy of winning young bird races at any price! Instead of looking to the future, the majority of fanciers are only concerned with the present.

You will never get young birds fit to race and fly long hours at times if you practice a policy of restricting their corn rations to mean amounts. Flying to the corn tin they call it. Systematic exercise coupled with a consistently calculated intake of food is of the greatest importance in the preparation of young birds for racing. I have referred before to Wally Austin who gave up the idea of training his young birds, as a result his youngsters owing to the regular periods of time they were encouraged to carry out were very well developed muscularly. Without muscle power you cannot ever expect your young birds to home in racing time. Wally Austin even contended that "in another 50 to 60 years, the homing incentive in the racing pigeon will be so much improved upon, owing in no small part to the severity of selection, and continuous racing, that in his opinion racing pigeons will not require directional training whatsoever!". My word I wonder what dear old Wally would say if he was alive now? Fred Pountney would have recalled the visit to Upper Walthamstow for Fred accompanied me on that very visit when we had the pleasure of handling Wally's top winning racers, including the Coronation Cup winner, and his two Fraserburgh winners.

My own opinion for what it is worth, is that for pigeons, young pigeons, that is, provided they really do fly well around home, you can take more liberties with them regarding the distances you jump them into races, than you can safely carry out with those that are not systematically given regular bouts of flight. Even so I would not advise fanciers to neglect road work. If you had the keenness to fly your young bird team both morning and evening, around home, after they had been trained up to say 25 miles, with a number of single, and double training tosses, you could be pleasantly surprised how well your pigeons would handle as the weeks go by. The finest way I can advise fanciers on the way to recognise physical fitness as it develops is to handle each one of your young birds individually before you release them for their exercise. What an enormous task for the fancier who breeds a 100 first-rounders! Fortunately the majority of ordinary fanciers on average breed 20 to 30 top number, so it is not an impossible task. But few fanciers are energetic enough to carry out such a task, preferring to open the loft and out they go.

Those who were prepared to experiment would be amazed at what they would learn from this method. Their youngsters would soon learn to accept such treatment. The timing of their flight begins when the last one of your team has been released. It is, believe me, a most enlightening exercise. You will the more easily realise what I have so often written about when I refer to pigeons that handle like brand new tennis balls! One must not lose sight of the fact that a greater number of fanciers are inclined to rely upon old lines in the family and because of this sentimentality continue to breed from pigeons that in far too many instances have failed to produce outstanding racers for several years.

Waiting for a good one to turn up now and again is not really the most wise policy to pursue. Winning pigeons are healthy pigeons, especially those that continue to win, but they are not easy to breed if you allow sentimentality to cloud your vision for the future. You have always to remember that you must be fully prepared to improve the standard of quality and type in your family. Selection for physical as well as racing qualities is of the utmost importance. If after several years you do not possess quality pigeons that have a true history of ancestral performances, either for yourself or for others you will not reproduce these qualities now or in the future. When you read as we all can if we have a mind to, of the loft reports, that abound particularly in the Racing Pigeon Pictorial, you will the more beneficially learn how very important ancestry of recent years is as opposed to what took place 40 or 50 years ago, despite the power of the genes that control the reproduction of the qualities we seek according to our desires and ambitions. Any form of breeding, whatever your ambitions may be, requires sound foundation lines, and in this respect the genes are of the utmost importance. You have to collect the genes that control the reproduction of the type of winning pigeons that you aspire to reproduce. You will as the old saying goes "not ever make a silk purse out of a sow's ear".

Late breds

Today as ever before there is still a great deal of speculation concerning the merits or demerits of late bred pigeons. Even all too often the definition of what constitutes a late bred is one that is not only bred too late to participate in young bird racing, but is also bred too late to complete a full renewal of its primary flights and of course its secondary flights. Many indeed are the numbers of fanciers who despise late bred pigeons. For many it is a matter of patience, yet also far too many fanciers are lazy and do not want to be bothered. There are of course other reasons. Space for one thing. But I suppose the main reason that many find it difficult to concentrate on pigeons once racing is over, and especially if their efforts during the OB season have proved unsuccessful. Whatever time of the year a pigeon is bred it is imperative that the subject is soundly constituted.

Any pigeon that does not look the part, lacks lustre, is mean looking, in plain terms looks a poor specimen can only reproduce pigeons of a similar type. Quality pigeons produce quality of feathers which is of the utmost importance, and pigeons that look the part, hence the saying already quoted above, "silk purse — sow's ear". Although fanciers of repute put forward sound reasons for not wishing to bother about the breeding of late breds, there are still a great number of successful fanciers that can also speak or write, both loudly or boldly as the case may be in favour of late breds. Many of the most successful Continental fanciers put themselves out, in order to reproduce late breds from their very best racers of the year. Yet for those with limited space, and consequently not a large number of outstanding performers to point at, there is no reason at all for them not to take a couple of pairs of late breds from their very best, even if they only have one outstanding performer.

They do not have to allow them to carry out the entire rearing of these specials. Think about it. That is the finest way you can collect the genes that you are looking for. Many a well bred desirable looking late bred has helped to bring success to those who have heeded such advice throughout the 100-year history of the racing pigeon. If you have a really good pigeon, that has shown by its determination and tenacity coupled with racing ability do not waste a single egg. Without being stupid reproduce all you can from such a pigeon, especially if it is a cock, without subjecting it to the rigours of rearing. With a hen the number is naturally restricted but even so you can make certain you never miss the opportunity to hatch her eggs. It does not matter how hard the foster parents work. What I am really trying to impress upon the younger fancier and those among us with less experience that good racers are not easily come by, so do not ever lose the opportunity to breed from them within reasonable common sense lines.

You may well ask, as many do, where have all the young birds gone? Who's catching them, and if they are, what are they doing with them? Most of those that are missing are either scratching a living in the fields around farmsteads, or are in other fanciers' lofts. They try hard to pass the buck by "taking them down the road" or "up the road" as the case may be. Some, because they like the look of a bird, test it out, and if it continues to train well, then report it in the hope that they may be allowed to keep it. And who is to blame them. Any stray bird that goes into another loft is more likely to settle and even race well to that loft, than to the loft where it was born. Probably the reason for this is that that particular pigeon has been treated better in his new found home, and consequently responds to VIP treatment! So instead of continuing to maintain the existing system of reporting strays, which by law must be claimed or else, reverse the law and notify the registered owner, that if he wants his pigeon back, he will have to claim it or otherwise he or she must signify that he or she does not require it and is pleased to enclose a transfer form and a few details of the bird's ancestry or even a pedigree. The revised Rule 67 of the RPRA goes a little way in this direction but not enough.

The returning of strays is becoming far too costly. On the other hand if a bird is wanted back, then you will send more than ample money to cover the expenses involved, which often total much higher than even the amount stipulated under RPRA rules. Many young birds are lost today because of the multiplicity of race programmes devised by the extremely large numbers of different organisations that now exist creates confusion owing to clashing en route that is inevitable with so many different organisations, including a large number of very small Feds creating different routes at far too many varying race points. I really am convinced that if young bird races were limited to 100-mile races only, whatever your Fed or Combine, young bird losses would be reduced by 50%. However, since the last time I advocated this idea, it appeared to fall on stoney ground, I now suggest that the first four young bird races do not exceed 100 miles, and that the new two races be flown from 150 miles, and the final race 200 miles or thereabouts. Any races beyond this distance would be catered for the Fancy by the NFC, NRCC and any other specialist club prepared to cater for such races.

Unfortunately the enormous numbers of young birds missing each year totally disheartens fanciers, and it most certainly discourages many who have by their own efforts and dedication discovered one or two, or even several good old bird racers, and maybe a good yearling or two, who because they are so disappointed with young bird losses, decide to stop further breeding for the year. In plain words they have had enough, when, in fact, they should consider the production of a few late breds from their better racers, and allow the less successful racers, or should I say returnees do all the work by rearing the "chosen few", bred from the "star" racers.

Such late breds with the likelihood of a brilliant "Indian summer" before us would enhance their loft's own future well being. These late breds could even be trained a few miles carefully during September and early October. Once the late breds from the best racers have been safely and efficiently reared you can then eliminate those feeders that have reared them. It is no use keeping those that continually arrive late, or are not prepared to race for you. Breeding only from your better performers, and annually eliminating all that do not, or will not race for you. This is a sound way of "improving the standard" as the world famous French fancier Pierre Dordin always stressed at every opportunity.

The problem of strays found by non-fanciers is always a much difficult matter and must be handled with great diplomacy. When it happens to me I really go out of my way to either collect the pigeon or find a fancier in the locality willing to help out. All too often I have other fanciers' strays delivered to me, or even collected by myself. Unfortunately most of these are not wanted, so have to be put down. It would not be wise to let the "finders" know about such action. What with neighbours who love cats, and non-fanciers going to such lengths to locate, or return birds to owners, a true pigeon fancier has to be a diplomat when he has to deal with members of the public.

Young bird losses are becoming a great burden for the sport to carry and we must do all we can to improve matters if possible. The most difficult problem of all is the weather. We are experiencing worsening and less consistency of the seasons. This phenomenon has been a feature of world weather for a long time now. Not for some time now the regular summers and pleasant early autumnal periods of yesteryear. These seasonal periods are only consistent in their inconsistency and as far as pigeon orientation is concerned are difficult for the racing pigeon. So much so that not only are many young birds being lost, but also old birds as many old bird competitors will confirm, especially those fanciers who had decided to switch their pigeons from North Road racing to South Road racing. These fanciers have been hit very hard indeed. Maybe if the RPRA were able to produce figures revealing the numbers of young birds reported this year to date as well as the locality by county of the finders and were able to produce figures of a similar nature for the previous year we might learn a little. It does appear that a great deal of the trouble stems from the very bad skies we are now experiencing.

Owing to cost our training methods have altered considerably. Years ago, before training transporters and private training schemes abounded, fanciers carried out their own training through the railways and by individual motor vehicle in very small groups. Also perhaps it is as well to be reminded that fewer pigeons were bred. The method of training one's pigeons today has taken the form of bulk training. It is not so individual today. Young pigeons are educated, at least many of them are, to fly in large numbers as opposed to the old days when most fanciers trained their own pigeons. Although I believe that many of today's young bird losses can be attributed to bad skies, I also reckon that the great deal less individual training increases the strayed, tired and lost brigade. Nonetheless I feel that it is possible for the Fancy through the help of the RPRA to collate a pattern of young bird finders and perhaps produce an information grid that may enable the Fancy to learn something about losses on a country-wide basis. We might

AUGUST

this way be more able to discover if the clashing is the major cause of these heavy losses.

Perhaps if we gave the matter a more detailed study it is possible we could learn a great deal more. There is no doubt at all that young bird losses today are as bad as they ever were and that racing generally, particularly for young birds, is not nearly so consistent for collective time of arrivals as it once used to be. Years gone by one could quite confidently look forward to closing up the loft within a half-hour or so of the time of the first arrival. Really, that is how young bird racing and indeed old bird racing in the early days used to be in the shorter events. Naturally as the distance extended old bird racing was then as it is now, for distance always has its limitations.

Leaden skies, dead skies, the type of sky that is associated with stillair conditions, humid conditions and skies without even a hint of a broken cloud cause far more losses than many would have us suppose. That is why it is so important to note such skies when one wants to go training. Don't go training with young pigeons at any time when the sky is dead — without a break. You can judge and please yourself when you train your own, but not so when you entrust your training to others. Young bird training is far more important than many realise.

In a really bad sky you don't have to go to the trouble of training to lose young pigeons, you can even lose them around home. The more consistently you train young pigeons the less likely you are to meet with heavy losses, provided always you observe and take heed of bad weather and poor skies. Some years it is extremely difficult to train pigeons as one should. Young birds that have not been well trained, nor experienced some form of individual training are more likely to be lost. Training, which after all is education, is imperative. When skies are low, filled in, without a break, or without a patch of blue anywhere, you are asking for trouble if you take chances in conditions of this nature. The humid conditions that have been a feature of the weather have proved disastrous to those who have taken too many chances.

Fanciers also should be concerned with the extreme heat created within the transporters when crates are packed to the legal maximum. Added to this is the lack of educating young pigeons to drink when confined to the crates for the first time. Young pigeons are very slow and often quite shy in learning to drink from the water troughs. Convoyers will confirm that often they consider it is a sheer waste of time and effort putting the drinkers on the crates in the early races. They report often that very few youngsters even attempt to put their heads out. Nonetheless most convoyers still go through the motions. However, for the benefit of the lesser-experienced fancier it should be pointed out that few pigeons will drink, even if they knew how to, from the inside of a crate unless they have been fed. At least this is true of young birds.

The prevailing conditions of transporting our own pigeons is overall a very costly affair, but it is here to stay. As a result economics have to be seriously considered, each journey has more or less to pay, and consequently the crates are packed as tightly as is reasonably possible. The pigeon has to be considered. There is not the slightest doubt that pigeons do travel well if they are comfortably packed. Comfort of the travellers is of paramount importance.

To my mind one of the problems we have to overcome at the race point is the normal method of supplying drinking water to our young pigeons within the loft. Mostly all pigeons (and we make no exception for young birds, at least not many seem to) are only used to habitually drinking water from a fountain, be it vitreous enamel (round circular water pans), galvanised, aluminium or plastic and these have formed the basis of supplying water for the pigeons within the loft for a long, long time.

Such drinkers (fountains) and there are a few other contrasting types as well, are a strange contrast to the water troughs now in general use at race points and hung on the outside of the crates or the baskets. The young bird crated for the first time is normally fed within the confines of his loft and among familiar loftmates, after being fed goes straight to the familiar fountain in the security of his home, and drinks from a water pan confined within a metal or plastic guard, that is crowned with a cone to protect the water from becoming fouled with excreta. A strange contrast indeed to a water trough hung on the outside of a pannier among a lot of strangers. What a nerve-wracking experience for these first timers especially if they have not been taught to drink or even feed from within a racing crate. There are fanciers who, having cared well enough for their pigeons at home, forget this most important part of their pigeon's curriculum. If fanciers do not take steps to educate their young birds to life within the confines of a crate it can take several races before they will have become wise to the need to drink a little, even if they have not been fed.

This is one of the two reasons fanciers of the old school, particularly when I was young, would always insist upon any age races. One was to help influence the younger pigeons to get a line under the guiding influence of the old birds. Secondly it was to initiate the novice young bird racers on how to survive in a basket. They copied their elders. When the weather was hot the old birds drank, and the youngsters would benefit from what they had seen. It all makes so much sense. In those days it was the famous railway pigeon coaches and baskets were stacked each side of the coach and staggered. Seldom more than two or three high. These type of races were highly successful.

It is foolish to feed pigeons for a short race, or any race for that matter, immediately before a liberation. Furthermore if the weather was normal not all old birds would drink unless they wanted to, and since quite a number do and will drink at the race point the drinkers were always put on the crates. Confined to crates or baskets, experienced pigeons will drink without a feed. They know the score and act accordingly. Not so the stubborn young bird racer that has never been educated to do so. It is the nature of the pigeon to drink in the manner that he does. No other bird in the world drinks in the draft-like manner of the pigeon except the sand grouse and even this specie comes within the order of the Columbiformes (it too drinks water in pigeon-style by sucking up its fill as does the racing pigeon).

If you were not aware of the unique manner by which pigeons drink then contrast it with the manner by which chickens, crows, sparrows and all other birds drink and you will then know the difference between scooping up water with their beaks as performed by all birds in the world compared to the sucking up method peculiar to all varieties of pigeon.

I really do believe it would help enormously in the early education of the young racing pigeon if fanciers and loft manufacturers were to incorporate provisions for the inclusion of water troughs as used at the race points in the young bird compartment. Several fanciers to whom I have suggested that they make a facsimile of their own Fed crate to educate their babies have expressed considerable approval after introducing the idea. For those DIY-minded enthusiasts they could easily make an extra side to which the water troughs are normally hung and then incorporate this item into the loft front without making it look an eyesore. Water your youngsters through this and I am certain you can visualise the advantages it would provide when they go to the races for the first time. Anything that will help your young birds to enjoy the fly home is something that we should all strive for. There are quite enough problems for pigeons to face be they young or old.

From reliable reports received from over a wide area, recent losses among young pigeons are probably greater than any season before. One fancier who commenced with 60 well-bred youngsters is now down to 20, and is considering retiring from further young bird racing in order to start next season with a team of 20 yearlings. This fancier, whom I have known for a number of years, breeds and rears exceptional babies, as I have witnessed in the past whenever he has called upon me en route for his initial training tosses. Another successful fancier whom I also know personally commenced racing with 52 well-bred and well-trained babies but is now down to 24. Another old friend and past winner of the Tommy Long Cup commenced the young bird season with 38 young ones and is now down to five and has decided to call it a day as far as the young bird season is concerned.

Another report concerns a fancier who entered 66 young birds in his first race and only got two or three back on the day, and at the time of the report three days after his personal disaster only then had a handful back. For many it is heartbreaking. My own fellow club members too are having a grim time. One of the most successful fliers locally, and who has trained well, despite many wins including at least fifteen 1st prizes in a season without any form of nomination, is also greatly puzzled by his startling losses. Others, also successful fanciers, that I have contacted over a very wide area have also suffered losses. Few if any of the fanciers that I have taken the trouble to phone or talk to are slack when it comes to the training of their youngsters both before and during racing, yet their losses have proved enormous.

A local club which promoted a very well-organised schedule of training tosses on a regular pre-race basis enjoyed splendid support. Consequently those regular supporters enjoyed excellent reurns at first and excellent results, including several good positions in the strong London NR Fed. Yet since racing has progressed some five to six weeks back losses have now greatly reduced the club entries and in some clubs considerably. In the first two races losses were negligible for those who supported their club's enterprising training scheme. It appears from reports that the main losses were among the members who had carried out only limited training, and some who had not even trained at all. Whilst the more annually consistently successful were gaining prizes including several very forward Fed diplomas. From a further survey since the second race, but particularly the third race, losses are now beginning to bring about a reduction in most club's total weekly entries. No more now is it a matter of serious losses among the lesser-trained pigeons but the losses are now becoming much more serious among consistently successful members. This is a situation that is reflected in many clubs.

One would have thought that in those clubs wherein a well-considered pre-race training programme had been initiated that losses would have been reduced to a very small percentage. But this is not the case. Doubtless with clubs promoting such training schedules they are overall assured in the earlier races that there are a far greater number of muscle-fit young birds than would otherwise be the case. As with all forms of bird life, and in this I am dealing only with their young, for in each and every species, their young birds have to become muscle-fit in order to be able to enjoy sustained flight. In this respect young racing pigeons are no exception. In fact, the muscular development of the wing is of paramount importance. In all cases, waterfowl, wild birds and racing pigeons, it is attained by easy stages. This is the real reason I advocate a gradual training programme. It is so essential for many too, it always has been since pigeon racing was first initiated over one hundred years ago!

If you have enjoyed the Open University television programme via Bristol and Birmingham Universities you might have seen a Canada Geese video which noted the importance of muscle-fitness and how slowly yet patiently this is achieved even by using a human foster parent. Most impressive viewing this proved to be for me. This is the reason I applaud wholeheartedly any kind of enterprise that enables fanciers to prepare their young birds for future racing aided by the facilities of gradually extended training flights. Muscle-fitness for flight is of the greatest importance in the early development of the young racing pigeon and should be encouraged at every level. However, despite club training and fanciers' own individual efforts that prevail among the most ardent young bird racers there are in my opinion far too many losses.

Another local club through one of its members' own enterprising efforts has also enjoyed excellent support for a privately run training programme. Furthermore it has been noted that those members who regularly supported the scheme have benefited considerably by annexing a number of Fed honours. Yet in spite of this local help, and after taking a "round robin" of fanciers' losses they are steadily becoming worse. The losses have reached enormous proportions including those by reputable fanciers of the greatest success, who report up to at least 50% losses since young bird racing began. There are some who contend losses are owing to the production of far to excitable and wild pigeons, as some would have it, too much or too close inbreeding. Although it may well be true that some families are more excitable, even wilder than others, I find this suggestion hard to accept in general terms. Others, too, consider it is owing to paramyxovirus and the vaccination policy advocated by the Ministry of Agriculture, Fisheries and Food. Concerning this latter I am not able to make up my mind. But of this I am more certain that if the policy of vaccination were strictly adhered to and carried out under controlled supervision with carefully recorded ring details, colour and sex of each pigeon vaccinated and maintained with "regimental discipline" then we might have been able to present a sound case for or against. Unfortunately the present lackadaisical approach to vaccination teaches us very little indeed, or so it would seem. Unless it be that more and more young birds are receiving the side effects of a mild form of paramyxovirus.

One thing is certain, far too many young birds are being lost. My own opinion is that far too many are being lost through the worst thing of all (since the railways were denied to the sport), that is too many race points in close proximity with far too many different race routes from north to south, south to north, east to west and routes on all other points of the compass. The result is that groups of stray young birds build up concentrations and attract exhausted young birds. Gregarious by nature young pigeons soon learn from others how best to fend for themselves. In time a few hundred work their way home, a few thousand are shot down, or fall victim to vermin, a few may take to the fields, live wild, and become semi-feral farm stock, whilst maybe a few become the progenitors of successful families through being reported and given to their finders. If all young pigeons that entered other fanciers' lofts as strays were initially offered to the finder who took the trouble to report to the RPRA, I really do believe that more pigeons would be immediately reported. Let the finder who goes to the trouble of reporting have the option to keep the bird as his or her priority reward for reporting. A pigeon that takes up residence in another loft as a young bird is far more likely to do well there than ever it will in its former home. Remember, I am referring only to young pigeons and not old birds.

Whatever may be the reason for these very high losses of young birds the custodians of the Fancy have a heavy responsibility. They must take steps to endeavour to improve the planning of race routes as well as reduce the amount of clashing that takes place. For the fanciers — disease, including respiratory problems, failure to check for coccidiosis, as well as a complete failure to treat for worms, are each in themselves causes for a build-up of missing birds, or very late arrivals. The facilities to remedy are prohibitively expensive.

An idea that I have put forward before, but one that seems to have fallen on stony ground is that no first races in any Fed should be beyond 60 miles. After at least three of these the next series of races should be 100 miles, with the final racing being 200 miles. Better still reduce all Fed young bird races to a maximum distance of 100 miles with not less than three at this stage. Remember, all of these races to be controlled by Feds only. Any races above this mileage would be organised on a provincial or National scale by specialist clubs. After a generally agreed recognised final Fed race date.

Failing this, longer races to be flown on a Sunday instead of the usual Saturday. At least the risk of clashing and confusion would be greatly reduced. Try it out on a National basis for a couple of seasons, and see how it works. Through the introduction of approved liberation sites and fewer race points it would at least be worth trying out. When it is a matter of 200-mile races, two or more local Feds could combine with one transporter, whilst the Combines would bring the young bird season to a close.

There are far too many fanciers leaving the sport. Of equal importance far too few young fanciers coming in. Our old ways were all right when we had our railways to carry pigeon traffic. Now that is all finished we must now do our utmost to encourage newcomers as well as the disillusioned to stay within our ranks. This is the reason

AUGUST

I have gone to such lengths to put forward ideas that may help us to curtail these awful losses of young birds. We certainly must try our hardest to reduce the enormous losses of young birds. If we try harder it may help to halt the serious decline in our ranks. It really is serious and will become even more alarming if we who have enjoyed a lifetime in this most fascinating hobby do not make an effort to figure out ways and means of bringing in and/or trying out ideas for improving upon our present methods for the benefit of the majority. Look around your local clubs and note the reduction in membership, and the complete absence of new faces, especially young faces.

My August chapter is almost at an end, and for the majority of fanciers who are after all hobbyists, breeding is now over. A few will not separate their pigeons, but doubtless the majority will. As indeed most fanciers have done for years; even as those before us did 100 years ago! Today there are different schools of thought concerning the separation of the sexes it is most certainly a simple way of reducing work. For those who have the right kind of nest boxes they can leave the cocks perching on the front of the nest box they have been allocaed from the beginning of the season. This method most certainly obviates a great deal of the problem of getting your cocks settled when the next breeding season comes round.

In this present day and age the majority of fanciers are only interested in the Widowhood method of racing and therefore the cocks that remain at the end of the season can be left in full occupation of their nest box that they occupied for the past racing season. It is best if the cock can perch or sit comfortably outside a closed nest box. On the other hand if you have ideas in obtaining, or already have Belgian type Widowhood boxes, then the existing male tenants of the Widowhood section will sleep and perch on a brick in the open half of his box with the nest side closed. Their influence is a great help next season when you bring in fresh occupants to the Widowhood cock section. Much in the same way with the Natural breeding sections. Especially as the majority have to introduce yearlings whether or not they fly Natural yearlings or Widow yearlings.

There are doubtless many advantages with the Widowhood system. Probably one of the greater advantages for those who rely upon the manufactured form of Widowhood fronts is that the actual size, as opposed to the old method of 19in x 14in Natural type nest box; the average size Widowhood box is 30 inches in length thus reducing the number of pigeons you are able to keep. This is the greatest advantage of all! Most fanciers keep far to many pairs.

However, if you are a Natural racer only, then it helps if you are able to allow your cocks to retain their present nest box. A shelf or perch on the front of each closed nest box will help you to settle your pairs next season. Always keep your cocks to their same numbered box, and as the next season soon comes round so it pays to plan ahead. If you are planning for new nest boxes then it will, I consider, help you enormously if you make individual nest box units. Size 19in x 14in is probably best, because you can still after all these years buy ready-made nest fronts of this size which are made in timber.The RP advert columns will provide you without much trouble. The size as I have often written about before is a legacy of the famous old type Tate & Lyle sugar box that were a special feature of many lofts countless years ago. When I was a lad you used to be able to buy these empty ready-made wooden boxes built specially by Tate & Lyle for sugar deliveries at 8d (4p) each! There was quite a loud hue and cry when grocers decided to mark them up at one shilling each. and I might add a greater hue and cry when they went up to 18d (9p)! They simply were a marvellous buy for pigeon fanciers.

There are a lot of new designs of boxes to be seen at The Old Comrades Show but for those who are planning to make, or purchase similar type boxes then I strongly urge you to consider about recessing the actual front five or six inches, preferably the latter thus enabling the cock to have a comfortable perch along the entire front of his box whilst separated. My own latest design are single units and are fully extended top, bottom and sides. It's great when you see how happy the cock is in control of his own nest box. Especially is this time more noticeable after his hen has been taken away. Always a sad moment when pairs are parted. My own Natural pairs are not separated until after Christmas. However, their boxes are fully closed. But with a shelf along the entire length of each nest box, the pair can perch comfortably, and are I consider much happier this way. It must also help them to moult without anxiety over the removal of their mate.

The exception to any form of breeding in September is if you are lucky to own a very exceptionally successful breeding pair. With regard to fanciers who are fortunate enough to possess an outstanding stock pair, then by all the means in your power you should not discard a late pair of eggs. Especially if the stock pair in question are getting on in years. The experienced breeder will know that for pigeons that are constitutionally sound, old age will not prevent them from producing sound healthy stock.

One can take common sense steps to prevent the valued stock pair from having to bear the full brunt of rearing, simply by the utilisation of foster parents. Good fosters are worth their weight in gold for they will hatch and rear the final eggs for the season. Always or at least whenever I am able to advise those fortunate enough to be in the happy position of knowing they have such jewels to hatch every egg possible. Not always is one so fortunate to discover the "golden couple". The success of a family of pigeons, is for the most part dependent entirely upon the emergence of outstanding breeders. For the most part, a single pair. The consistency of the reproduction of the "elite" racer is dependent upon these naturally endowed reproducers. Looking back over the years I regret very much that I failed to take the opportunity to hatch every egg laid out from a proven stock pair, or every egg sired by a known stock cock.

For those who possess a really outstanding male bird my sincere advice is that they take every opportunity to mate such a pigeon to as many good hens as they possibly can. If such a policy had been carried out by those like myself who have kept pigeons for more than 70 years we would have done the Fancy a major benefit. Again in retrospect I regret very much that I let a number of such opportunities slip by. Breeding from too many untried pigeons, and unproven pairs is both foolish and a waste of time and hard earned money. It is far better to breed a greater number of pigeons from the proven breeders and utilise the unproven to carry out the duties of foster parents.

I learned from the 1989 King George V Challenge Cup winner from Pau, bred and raced by David Delea of Dagenham, Essex, that his winning pigeon is a descendant of a famous pigeon known to old-time fanciers as Champion Major, a black pied cock, black splashed head, all black body with white flights, and slight white socks, ringed NURP26318. Champion Major was an outstanding looking pigeon and was bred and raced by R W Beard of Kenley whose very successful family were based upon the old Barkers. In fact, were always described by Mr Beard as the original Barkers. However, it is possible that most other strains were used from time to time, like most fanciers always seeking to improve upon their existing stock. If you look back through old publications black pieds were not a usual feature of the early N Barkers. However, there is not the slightest doubt that Champion Major was a top class breeder.

Furthermore Champion Major was always included as a descendant of the original old Barkers. He too was a consistent racer as his records before me as I type these lines confirm. He was probably unraced as a young bird. In 1927 he won 1st prize Dol in the old Croydon Wednesday HS (my father and I were members of that club, and in that year too). In 1928 the then little known un-named black pied won 1st Leamington (*I am almost certain this should be spelt Lymington—FWSH*) in Woodside CCS, and 1st Crystal Palace Fed (2,790 birds), flew Bournemouth winning all pools, 2nd prize Jersey "News of the World" Championship, all sections. A race promoted by the Sunday paper of that name for 1st prize-winners only, with stipulated numbers laid down. In 1929 6th Leamington, 1st Bournemouth Croydon HS, and flew the La Teste smash, many birds missing. In 1930 2nd Bournemouth, 7th Crystal Palace Fed, 2,352 birds, flew Laval and Marennes winning in both these HP Nationals. 1931, 2nd Dorchester, beaten on trapping by the noted Kenley Blue.

In 1932 3rd Weymouth, C Wed HS, arrived before Mr Beard's club winner, 5th Bournemouth Woodside HS although he actually dropped with Mr Beard's Fed winner known as Beard's 547.

By this time Champion Major had developed a shyness to trap. In 1933 Champion Major won 1st prize Fed from Dorchester by a margin of 60 ypm. He had by then flown in English Channel races 14 times, including two smashes, when many splendid pigeons were lost. During these racing years Mr Beard's black pied gained a reputation as a breeder and it was because of his breeding ability that he was given the title of Champion Major. Besides winning the Fed twice, he was the sire of 11 different Fed 1st prize- winners. The number of his grandchildren proved numerous and included a great couple in 209 which won 1st prize Berwick, 3rd NR Fed and also flew four times San Sebastian, twice lifted from Berwick into San Sebastian, Spain, and in 1927 won 18th Sect, 33rd Open Mirande National FC; the other was 410, flew and won 1st London Social Circle Mirande, 25th Sect and 145th Open Mirande National FC and was awarded RP Tankard and a "Peoples" canteen of cutlery.

I do recall my visit to Kenley with the well known L Gilbert, a past secretary of National FC, and later president of the London Social Circle. The famous Sherwood Mount Lofts literally won hundreds of prizes. Unfortunately I have never seen a written pedigree of Champion Major, but it was published that his family of pigeons were based upon the original Barkers. It has been suggested that he was of the old Baird of Liverpool Soffle family, but I know for certain that this is not true. In the passage of time the written word can sometimes become distorted, or misinterpreted, ie as took place in the 1914-18 war when signals sent out a message "send reinforcements, we are going to advance", and by the time it reached its destination became "send three and fourpence, we are going to a dance"! However, Champion Major did make his mark for he became grandsire to three National winners as well as grandsire of the winners in 1939 of the Three-Bird Average Cup 600 miles in the 1939 National Flying Club. It undoubtedly was his ability as a producer that brought him enormous popularity and National fame. And to add to my strongly held conviction that you cannot ever deny the great producers for they will continue to prevail (and well proven if records can be researched), and as has been revealed by David Delea in his winning the 1989 Pau Grand National with a descendant of the immortal black pied cock, Champion Major. It is most pleasing for the writer to be able to have the opportunity to record another King's Cup winner that is descended down through several generations of British bred long distance winning stock.

SEPTEMBER

YB Education — The Litchfield Crate Clashing — Transporter Crates — Holbard & Son — Filling Eggs — The 1986 Rome Race

September sees the end of young bird racing, and so another season of racing has come and gone. Many young birds have been bred and raced. Thousands in fact! Many hundreds, thousands in fact have been lost. Many are bred that should not be bred. Many are lost that should not have been lost. Among those lost are those that we can ill afford to lose. Among the lost are probable future breeders. Outstanding producers are a rare commodity. Although not all pigeons lost are good ones not by a long chalk! Sometimes these losses are due to our own desire to win just another young bird race, or a special produce prize. All too often simply to win a young bird average trophy. You can take the pitcher to the well once too often.

Yet another valid reason for losses is because the majority of fanciers breed from all their pairs when it would be wise to breed from only the very best pairs. Let those that you feel are of lesser breeding potential carry out the work of feeders — or foster parents as they would become. If you find good feeders they are indeed invaluable. If you are fortunate enough to own a proven stock pair, or perhaps a proven stock bird be it cock or a hen, then make certain you hatch every egg they are responsible for. This will strengthen your hand for the future. At the same time it will also give you the opportunity to use the pigeons less rated as feeders, which is after all an important duty. But only if you make absolutely certain that the babies are never left unfed. When this happens and if too often fretted flights, both tail and primaries will become a disfigurement in your babies.

Although it is not so easy to pinpoint all the reasons for the enormous losses of young birds and there are several I am convinced more than ever that the majority of losses, particularly in the early races, is owing to clashing at the race points as well clashing en route. Many of those missing run into trouble at the race point. many others for want of adequate training are automatically late, or get lost. They become confused, lost and weary, they find a haven for food and water and thus learn early how to survive by living rough. Many go into other lofts. Many of these early race latecomers return days, weeks or even months later. A few lucky ones may even become good pigeons later even good yearlings. One can always forgive a well bred pigeon that comes home of its own accord. This also proves my point that not all pigeons lost are bad ones. There are many other reasons why pigeons go astray. The weather too plays a great part. So too do bad liberations.

Fanciers too are to blame especially if your young birds have not been educated to take a drink before they are released. In this respect any heatwave will find the weak link in our armour. Among the reasons is the all too often complete lack of regular training before racing proper commences. Training young pigeons is both exacting as well as expensive. But you must be thorough — and you must observe the weather. Probably one of, if not the most important reason of all, for early season losses is a complete lack of thoroughly educating your young birds (discipline probably is a better word) to drink from the baskets, or the crates. My repetition is intended because in my opinion it is so important. More so than many realise these early races thousands lack crate or basket training and young pigeons quite unable to take a drink when at the race point. They simply have never been educated or drilled on how to survive, many are too timid, others all too easily intimidated. This is undoubtedly the major cause of losses and late arrivals from among the early entries. Many of the missing go down for water, or something that looks like water with disastrous consequences.

My own view is that we should consider the incorporation of an extension to the loft that takes the form of a crate designed exactly like the crates used by your own Fed. They could even be fitted under the traps or Sputniks. You could, if organised, at Fed level, purchase these crates from those responsible for making the Fed's crates. Fit this in such a way to the young bird loft, or the young bird section. This is the best idea that I can put forward now. If it is given careful thought it will enable you to educate your babies from the moment they take their very first drink. Especially if you make certain they need to quench their thirst when first introduced. Try to imagine how helpful it would be if all lofts incorporated a drinking parlour in the form of a basket or crate in the young bird section or loft.

Imagine too, how very effective it would be in the early education of young pigeons. Such a fitment built-in would I am certain solve, or at least help to alleviate the discomfort or suffering of young pigeons being liberated for the very first time. It must be an unnerving experience to find themselves in the company of several thousand pigeons and strange crate or basket mates without knowing how to quench their thirst before release. My suggestion, after much thought is that from their very early days of entry into the young bird loft, or section, the crate-like fitment would ensure that their drinking would be from within a Fed replica fitting. Think about this simple innovation before next year's young bird season comes round.

Overall I believe we need to pool our thoughts and ideas in order to prevent the continuance of the many early season losses the Fancy suffers. Just a thought, it would be better still if the design is the same as those used by your own Fed. When ordering a new loft, or making one yourself, make certain such a fixture could be fitted. If you are clever with practical designs share it with other fanciers. The ideal would be if all fanciers got together through their variuous Unions and/or Feds with the idea of producing a standard transporter. It this were possible and I see no reason why this cannot be achieved in due time, we would then discuss the introduction of a standard-type crate. Although costly there is plenty of room for improvements in the designing of transporters.

In order to implement effectively such a revolutionary idea it will be necessary for fanciers to work toward a standard-type crate. One of the finest crates that I ever saw as well as used was one designed by the late Harry Litchfield of Heckmondwike, a nephew of the famous Bill Cooling of Wakefield, Yorks. Both were formidable fanciers. Both established unique families of pigeons. There must still be many of their descendants in and around Yorkshire even now. Pigeons like these two great fanciers cultivated simply to not disappear altogether without a trace! The foundation of the Cooling family was based on the N Barkers, the James Moss, and the Belgium strain of Bosack. The Litchfield family were based on three cocks that were, in fact, bred by Bill Cooling and each of these three contained the Bosack strain. This strain was very successful despite the few only that came to the United Kingdom.

However, I am getting away from the point at issue, but I do know that both uncle and nephew were always ready to discuss young bird losses and that is why Harry Litchfield himself spent so much time and money experimenting with various designs in transporter crates. I will give an outline of the crate and its special features so that fanciers may commence to think about the idea of a standard-type crate that can also be incorporated into your lofts when, as many will, be considering alterations and improvements before the real winter hits us "good and proper".

The Litchfield crate was made of top quality eight-ply. Its main feature was the prevention of 'bullies' terrifying the less bold to obtain a drink. They included removeable false wire floors which prevent any pigeon in each crate from eating corn that had been regurgitated inside the crates, which is often noted at race marking. The wire was strong. These Litchfield crates fitted into the vehicles longways in, and not as the majority do wideways out. I hope you can understand this. When pigeons were released, "they flew out in waves", to quote a fellow Yorkshire fancier. The wire floors impressed me considerably. Probably there may be some still in use today. Perhaps some kind soul will drop a line to me.

From the reports of the final races be they Nationals, Combines or Classics the total numbers combined in these events be they North to South, South to North or East to West, indeed bear in actual total, no comparison to the total of rings sold. But at least those survivors are the owner's hope for the future. Survival is what it is all about. However, equally I am also certain that owing to our slapdash methods we lose many birds that, given the chance, would prove themselves valuable assets for the future progress of our aims and ambitions. As I wrote earlier many are lost that should not have been lost. However, thank goodness at least that the record entries reported in these races, coupled with the numbers reported home proved that these end of season competing birds had sufficient education to enable them to survive the season's programme. And thus we shall learn in due time, throughout the winter months how good some of those who excelled were bred and subsequently trained and managed by their proud owners. Gladly I note that several among the successful were old established and successful old timers. Equally it is pleasing to note the success of the younger fanciers.

One in particular is Mick Jarvis Jnr, who prior to moving to Hutton, in Westside, Essex flew with his father Mick Jarvis Snr in the old established Enfield Excelsior & Borough FC. In 1990 Mick Jnr won 1st, 2nd, 5th & 7th Open Essex Combine from Morpeth. This well conducted organisation attracted almost 3,500 birds for their final race. I have known Mick Jnr since his very early teens so was equally thrilled to hear from my young friend. The work he has put into these young birds has been truly enormous. Married, he and his wife Julie, a daughter of Pat Newell, formerly of Wood Green and now a resident of Billericay, have two sons Matthew and Daniel. When I asked Mick had he given his Verheye Busschaerts a name his reply was: "No not yet, perhaps you will do this for me", always a tall order, especially when I am asked to devise a name that incorporated his complete family name, his own, his wife's and their two sons. As a result and some thought I have named his winner Mijumada, which incorporates the first two letters of Michael, Julie, Matthew and Daniel.

Both for training and racing YBs there are many who complain of losses, many indeed report very large numbers lost and not all are novices either! When you hear, as I have, of the very top fliers reporting heavy losses openly surely this indicates that all is not as it should be for YB racing. For several seasons now YB racing and training has proved a costly business in terms of losses. The extreme heat we have experienced of late, and especially in the last several years, leads one to try to figure out how these losses can be reduced. It is indeed a tall order to venture into the realm of speculation in the hope that a solution can be found, nonetheless we have a duty and therefore must try. Clashing is, in my opinion, a major contribution and this has undoubtedly increased enormously since we lost the use of the railways and transporters to races, and of course for training purposes.

Try to visualise how many different liberations take place every Saturday, then think also of the many scores of varying midweek races that take place. Against these colossal numbers of young pigeons of varying ages involved, many of which do not arrive home on the day of release, many not at all, and for various reasons — either owing to accidents or through "mistaken identity" by so called sportsmen with trigger happy fingers. Others are lost to predatory birds, but many others are missing due to lack of sufficient education, ie without crate or basket training for both feed (holdovers) and especially water training after being confined to a crate or basket.

I have long advocated the incorporation of a crate-like bay, or the facility of having a panel in your young bird section that resembled the side of a crate, preferably a reproduction of a side as used in your own Fed's crates. Unfortunately we, the fanciers who make up the sport, have not got round to the idea of making a standard-type crate. What a boon this would be to the sport in general.

There is no doubt a great deal more that could be carried out to improve the conditions for young bird racers. Despite the losses that occur for the various reasons I have put forward, these do not of themselves account for the many youngsters missing, many find a better home at a well endowed farm. First far too many youngsters are bred from parents that have been forgiven for several mistakes in their careers. Secondly, because of the numbers reared in the first round, coupled with the fear of losses based on previous years, a second round is then reared to offset heavy losses expected! Yet no consideration whatsoever is given to the size of the average fanciers' YB loft, or as often applies YB section. Thus you have overcrowding at its worst. This is where your troubles begin.

Overcrowding leads to fighting. Such upheavals among overcrowded occupants brings about its own problems. All this discomfort is also responsible for the exacerbation of colds and respiratory problems which, if not noticed even for a day, will bring about an increase of virulence in the course of a disease. I does not take long to establish respiratory problems in an overcrowded loft (pen) of racing pigeons. Respiration is of such importance yet quite seriously overlooked when we plan for YB racing. Whether the planning by the individual fancier with his limited facilities of keeping pigeons without the necessary space for an increase of loft numbers when young bird breeding arises, or the organisations (Feds) who are, perforce, obliged for economic reasons to insist the crates be packed to the full in order to accommodate the very large numbers of youngsters they have to cater for in the early part of the young bird season. Think about it! Not enough room, and therefore restricted breathing space in the YB loft, and consequently owing to increased numbers, an overcrowding of the crates en route to the race points.

It should not take much imagination to visualise the tremendous build-up of heat in weather that is already humid, hot or very hot! Such conditions, where allowed to prevail, encourage "carriers" of disease. They pass by undetected. In no time at all you are faced with a problem of epidemic proportions. Too many pigeons kept brings about its own problems and the result, therefore, is you have a situation where far too many young pigeons are not fit enough to be allowed in the race team, let alone be entered for the races.

Many YBs suffer a wing or leg knock and if not recognised the victim is severely penalised. Television aerials account for a large number of accidents, so do wires, glaring sun and clear skies and no cloud cover. You have to be most vigilant at all times. Once you note a YB that is loathe to fly, keep it in, and above all carry out a most exacting examination. Realise that a bruised wing or damaged foot, no matter how slight, is always a severe handicap to a creature that is dependent entirely on its wings and feet (undercarriage) to satisfy your own ego to clock in the quickest possible time. Similarly an athlete cannot run or jump or throw a hammer or javelin with a bruised foot or torn muscle.

Now that Global Warming has been established due to Man's ignorance of the problems which had brought it about in the beginning, by creating a very enormous hole in the ozone layer, we pigeon fanciers will have to try a great deal harder to make our pigeons' task as easy as we are able. That does mean that we really have a great responsibility to ensure that our feathered friends are given every possible help. It also requires a great deal of money. Advanced design of transporters far and away better than ever before. Ensure that crates are well designed and fitted with false wire bottoms to avoid possible contaminated corn that has been vomited from being consumed by the hungrier contestants. Corn is brought up often due to nervousness, largely because pigeons are fed far too late before being basketed when taken to the race marking. But vomited corn can also lead to an infection contagion. First-time racing is an ordeal, especially so for the very young pigeon that all too often is ill-prepared for a race. Remember too that pigeons seldom drink naturally until they have been fed. If you feed them at the race point as soon as they arrive in order to get them to drink afterwards, they would have lift-off problems like an overloaded cargo plane, and most certainly would not be inclined to trap "pronto" at the home end!

There is no doubt at all that YB racing can, and does, bring about the loss of very good pigeons. Not all young pigeons lost are lacking in ability and ancestry. No wonder many continentals do not race YBs at all. In these days of Global Warming and consequently hotter months of July and August, it would be wise to at least duplicate, better to triplicate early race points and keep them from around 60 miles to 90 to 100 miles. The seasons have changed or are changing. Heatwave conditions, glaring sun and no cloud cover produce problems for OBs with lots of experience to fall back on. How much more difficult must it be for mere babies? Far too many race programmes commence with a first race of 100 miles or more which it really is a long way for youngsters. Not satisfied with a repeat, the distance of the next race and those after increase dramatically. The policy at the present time by the majority of Feds is too severe for YBs. Well-bred youngsters do not have to be thrashed with too many long distance races. For some years now I have wondered why Feds do not confine the races to several at 60, 80 or 100 miles, one at 150 miles and a final one at 200 miles. I really do believe it would benefit both fanciers and the young pigeons for the future.

As I have already implied there are far too many different points of liberation taking place all over the British Isles on the same day. Among these vast numbers of pigeons, one could find injured pigeons, as well as those that have been previously injured but passed unnoticed. And of course many duck-hearted pigeons bred from partners who have not been pardoned for the many mistakes they have made in races and training tossses. May I also remind all concerned that the width of the country is so comparatively small that according to wind direction — westerly or easterly — a pigeon can, and in numbers too, be off its route in no time at all. The build-up of stray pigeons on farms and farmlands is quite large in parts, and very influential in encouraging others to join them, especially the duck-hearted. We are looking for outstanding pigeons, champions in fact, and such types are indeed rare, particularly when the distance increases. Good average pigeons, and indeed better than average racers is the target you must set yourself if you are to succeed as a true long distance pigeon fancier. When the weather is fine, meaning fine for those stray pigeons content to sit about gorging themselves with seeds and other tasty bits and pieces to be found at this time of year in the countryside, growing fat and lazier, you are best off without their company. In fact, they are best suited to those with culinary talents! Unfortunately some do make their way back when the weather gets more wintry and encourages one into believing they are worth breeding from. Be very wary of them all.

Young birds that really do have ancestry can be given a second chance but no more than that! Although, as I know from experience, you can sometimes discover a good one that has worked his or her way home after a long absence. Indeed, to justify my pre-observations that not all those lost are all bad ones! You can and will find the exception to the rule but they are not all that plentiful.

Years ago fanciers used to breed their young bird team by mating up mid-March and then more or less stop all further breeding. At least this was true of many fanciers. Today this is far from the general practice. Thousands of youngsters are now bred very early in January. Such stock are really late breds from last year.

There is not the slightest doubt at all that many thousands of young birds are bred each year solely for young bird racing. At present young bird racing is more popular than ever. In days gone by pigeon racing was for the majority, a matter of old bird racing, followed by the normal young bird racing programme with the final Fed race providing the final test for the year so far as club racing was concerned. The Combine Classics and the Young Bird Nationals have now become the Grand Finale for the majority. Whilst for a smaller minority the Open young bird races, that now often follow the Fed racing programmes have become most popular. But that's not all, now we have the Great Futurity races as well. As a consequence of the enormous amount of young bird races of prestigeous importance now promoted, more young birds than ever are being bred.

Further to this we are consequently producing a tremendous number of young birds each year all of various ages. Unfortunately in adddition to clashing en route we are now producing a further clashing within the ranks of our youngsters, meaning the clashing of ages within our lofts. This would not be so serious if we were able to find the time, and be able to provide the facilities to keep each age group separate; both for the exercising and intitial training. Since this for the majority is impossible we resultingly create confusion in the minds of out immature own bred stock. Try to imagine the great confusion that would be created within the minds of children if each grade among the very young were all graded as one class! Young pigeons are like children, most impressionable. They will imitate the adult pigeon. Gregarious by nature the young ones will follow the larger groups. This is the reason that for years now I have maintained that soon

SEPTEMBER

after youngsters commence to "go on the run" daily it is essential that young bird training begins within a matter of two weeks from the first time you note that your babies are on the run, never mind that old bird racing is in progress.

Far too many fanciers neglect considering the commencement of the training of their young birds until old bird racing is finished. Once you have schooled your young birds you have reached a stage whereby they will have the knowledge to work back if they become mingled with other pigeons. I do not believe I am alone in this theory, there are a number of fanciers who think this way by acting accordingly. Since it is sometimes difficult to convince fanciers that some of the ideas and theories advanced through these chapters may be only the figment of the writer's imagination and not based upon experience of a personal nature, I decided to approach one of the most successful fanciers and endeavor to obtain his ideas on this very question.

The fancier is Freddie Holbard junior of the formidable Mr & Mrs Holbard partnership. I felt confident that young Freddy would have given a great deal of thought to such matters — and I was right. The 1987 LNR Combine Young Bird Classic winner delighted as he was that he and his wife Denise had at last realised a great ambition, having previously been 2nd Open Combine, is indeed a very studious type of fancier. Much thought has been given to previous mistakes, for as young Freddy remarked, we all make mistakes.

His own young birds were treated well between the Stonehaven and Thurso LNR Combine races. Freddy reasoned that it is a grave mistake to allow one's young birds especially those bred early to range continually without purpose. Training is therefore introduced into their curriculum soon after they have commenced to "go on the run". There is, states Freddy, the serious problem of clashing en route and also the clashing that takes place at the race point. There is also, says the London NR ace, the present day trend for each and every one of us to breed far too many young birds. With this I could not agree more. Of great interest to me was Freddy Holbard's observation regarding the number of different age groups produced by many fanciers within their own loft. This fully endorsed my own thoughts about the "age clashing" within one's loft that all too often leads to confusion among the youngsters and subsequent losses.

Freddy also believes that owing to the prices paid today for new stock, or even squeakers, that far too many pigeons are patched up with various kinds of "pills and potions" because of the value placed upon them as a result of an expensive purchase. These were not exactly his words, but more or less what Freddy implied. A most thoughtful and dedicated fancier he added, "It is all too easy for older fanciers to observe when asked about young bird losses, as well as old birds that have been placed among the missing, to shrug their shoulders and suggest the real cause is because 'there is something in the atmosphere". It was refreshing to note that without any reference to young bird ages, the time to mate or any other aspect of this much debated subject Freddy Holbard junior was most emphatic in his condemnation of too many age groups within a single loft.

Another frequent statement he has often made in my presence is the importance that education plays in the development and management of a loft of young birds. "Keeping and breeding too many young birds is a mistake for you are not able to attend to their early education in the manner that one should. Education for young birds is very important". To substantiate this observation Freddy added, "My own team of youngsters, 25 in total, were well schooled soon after they commenced to 'go on the run' and after a highly competitive season I finished up with 17 out of the total 25 I started with."

Young Freddy's father, that great character Freddy Holbard senior, also won London NR Combine with his famous red pied hen from Fraserburgh in 1956 beating 6,599 birds. Is it a unique record, two members of the same family winning the Combine? If you include Brian Haley of Cheshunt young Freddie's father-in-law who won two Combine races with his Glamis Happy Girl and Glamis Solitaire, it creates a further unparalleled achievement within a single family. Denise, the Morpeth Combine winner, named after young Freddy's wife is bred from the noted Bonecrusher which was bred by Freddy from the Brian Haley dark chequer cock Sonny bred in 1984 and which sired, until the beginning of the present season, several very fine races. Dam of Bonecrusher known as Don't Look, is a double grand-daughter of Champion Shy Lass has in addition to Bonecrusher produced The Guvnor winner of four 1sts; 15th Sect, 91st Open LNRC Berwick; also Just Look 93rd Open LNR Combine Morpeth; and Young Look a very promising breeding hen. Dam of Denise is the blue hen, Just Right, which young Freddy told me was his father-in-law's choice when the two visited the 1984 Old Comrades Show and attended the Auction sale.

This hen was given to the Old Comrades Auction by its breeder Alan Britton of Corby, Northants. Her breeding is sire Darkie, bred 1976 winner of five 1sts in club and three times 2nd in Open races, whilst the dam was bred by Ron Hallam.

There are several unique features of this 1987 London NR Classic race from Morpeth, not only the record of a father and son success in winning the Combine in their own right, there is also the unique record of four Combine wins for a family united in marriage but also as observed by Denise Holbard is the coincidence that besides the first three Open prize-winners all being Busschaerts the names of the winning owners all begin with the letter "H"! There is I am certain a great deal more that could be extracted from this LNRC young bird

SEPTEMBER

Derby but I feel it must wait for another time, or even be reported by others. However, I cannot let this pass by without also recording the fact that could escape many and that it is ironic that an ardent Widowhood specialist like young Freddy Holbard junior now is, should win the Combine with a hen!

September is almost over, and another season of breeding and racing has been completed. Great honours have been obtained by the few, and many have been rewarded for their indefatigable efforts. For the many it has been a year of trials, even tribulation, but then this has always been so for the past 100 years! Nonetheless for the real fancier who has an almost indefinable love for the fascinations of the racing pigeon it has all been more than worthwhile. In the sport of, or as I prefer, the hobby of racing pigeons, the richness of the rewards are in the production of a true athlete, a pigeon that has the toughness to endure in spite of the power of nature to change almost without warning the weather and strength of wind force en route from the race point to the loft however modest its home may be.

With good management you may, by ensuring a happy atmosphere to the newcomer and a suitable environment, help the 'old stager' to fit in to its new home without too much trouble. However, not all pigeons will accept a new home after ten or more years in the loft in which they were bred. I received from a very old friend a famous racer that I have always admired. Not only for its long distance performances but also its physique. It was a first rate racer and undoubtedly one of the best racers my friend bred. Now in its 16th year the cock shows remarkable sense. He is quite fearless of my presence. His physical condition also indicates the possibility that given the right conditions and the hen that pleases him most, it may be possible to get a fertile egg or two from him in his new home. But we will have to wait and see, and also have lots of patience. He enjoys a chat with me, as I do all the talking! Good foods, vitamins, wheatgerm oil and sprouting peas will all help.

Importantly, this old pigeon must have above all else his own domain, where he can remain with the hen of his choice, which I might add he thinks he has already. In the first place he was placed in his own loft with three highly regarded hens of his own ancestry. The three hens were separated by a dowelled partition which allowed him good viewing. Having made his choice, that hen was introduced with absolute success. The hens I selected each possessed wonderful temperaments. Unfortunately to date the two pairs of eggs laid have both proved infertile.

Although the response to my appeal for information on the breeding of all the arrivals from Rome 1986 was disappointing I have received news from Australia concerning the breeding of the pair of pigeons that were presented to the late Glyn Parry of Orpington, Kent and from which the first recorded Rome arrival descends. My Australian correspondent is none other than Bill Button, late of Ipswich, and now living at Salisbury Heights, S Australia, where Bill flies in partnership with another Pommie, Harry Nicklin who previously hailed from Dudley, W Midlands and who used to race in partnership with his grandfather in England. The partnership has not done too badly this their latest win at the time of writing being from Worth YB Derby (360 miles) where they won 1st Club, 2nd Sect, 6th Fed, 11th Open with 8,000 birds competing. However, Bill's correspondence related to a letter he received from his old pal way back in Army Pigeon Service days, John Crosland of Halifax. John wrote and told him of my reference to the first arrival, a silver blue hen, from Rome that was reported by Harry Kennett of Orpington. The pigeon, which reduced the journey from Rome by around 18 days, was described by Harry Kennett as descending from the Bill Button family of pigeons. Doubtless Queen of Rome contained more than a sprinkling of the Button family since her colour marked indelibly that characteristic.

It all began in 1972 when Bill Button presented George Parry (in the Army Pigeon Service the late Glyn Parry was always known as George) with two squeakers, 72P22607 and 22609. Colours of these two pigeons were not recorded in Bill's books. Fortunately all available breeding details were, and as Bill took with him to Australia all his old Squills and other books he has been able to submit considerable information of the original presentation pair of squeakers. It makes interesting reading and confirms again my opinion that well bred pigeons from proven lines live on through their descendants if only fanciers have the patience and energy to maintain their annual breeding records. Anyhow 72.22607 was sired by blue cock 69S28401 known as the Stow Cock that was bred by E Wright of Stowmarket whom I met and even probably sold pigeons for, at the old London Auctions many years ago.

This fancier flew tremendously well, particularly at the distance, and the Stow Cock won at least one 1st prize for friend Button, and he was bred direct from two of E Wright's East Midland Centre Gold Medal winners. These coveted medals were won for their performances from Lerwick and Thurso. Dam of 22607 was a blue chequer 67S41479 which was a line bred daughter of a pigeon named Evening Star 63A65142 when he was mated to a silver blue hen 65D62240 known in the Button records as The Belle. Evening Star was from an unrung blue chequer pied cock son of 936 and 943, a silver pied hen, whilst dam of Evening Star was 938, daughter of 1240 and 336. The Belle 65D62240 was bred from 368 and 9863. The former, 368, was bred from Evening Star when he was mated to a silver blue 65151 who was bred from 2251 and the hen 943, a silver pied hen, and the grand-dam paternally of Evening Star. The breeding of 22607 was

SEPTEMBER

maternally all Bill Button's old Ipswich family. If I had more time I am certain I could trace the ancestry of 41479 back through the years, that would make most interesting reading for those (alas few in number) who have the desire or the patience to study what is one of the most fascinating sides of the hobby, long distance pigeons.

In running on in my enthusiasm I almost forgot to include the dam of the silver blue hen The Belle, ring numbered 9863 which was in turn a daughter of 936 and 934, and note 936 was the grandsire of 63A65142 who was Evening Star. Both Evening Star and The Belle were prize-winners. They were undoubtedly both greatly prized by Bill Button, for indeed these were the two pigeons selected by Bill to accompany newspaper reporters who reported upon the entire British Army manoeuvres on Salisbury Plain and these two pigeons homed with message containers filled with films of the British Army manoeuvres of that time. A photo of The Belle hung in the Old Belle Hotel at Stoke Bridge, Ipswich. Evening Star was sent to Lerwick and was reported in South Germany after the race and transferred to the German fancier. However, Evening Star was not satisfied with his new-found home and returned to his beloved lofts in Bill Button's garden the following January having crossed the North Sea during a very dreary winter.

Further information reveals that 943, the grand-dam of Evening Star, a silver chequer pied, a most important link in the Button family, was bred by Bill Alty of Ipswich from a pair of birds that Bill Alty had from Bill Hammond, of Hammond & Whittaker, no doubt of Grooter bloodlines. Grandsire of The Belle was that great-hearted pigeon, Evening Star which showed that he was not afraid of flying over strange land or sea, and what is more proved capable when fit of tackling the long journey from Germany in the middle of winter. Again students of breeding will note the value of close inbreeding to the silvers in the family, as well as the inbreeding to Evening Star.

The other Bill Button presentation was to ex-Army Pigeon Serviceman the late George Parry. Ringed 72.22609 it was bred by Bill from dark chequer cock 69F4059 and a blue chequer white flighted hen ringed 69F4058. Both were of outstanding breeding to the best performers, and again an example of inbreeding to both the best breeders as well as the best performers within the Button family. Consequently these were inbred further to the silver line. This no doubt is the reason that fanciers of many years ago were wont to express satisfaction whenever they produced a colour that was the facsimile of a previous good performer, or breeder as somewhat more rarely happens to a pigeon of a colour that was both a good racer as well as an exceptional producer. It does and can happen but it is the exception. Not always are the best racers the best breeders, often it is the children that excel as breeders. However, sire of 4059 was Bill

Button's noted racer Uno Solo, a blue chequer cock 66B82793, winner of several 1st prizes and other good prizes including 1st Ipswich Open Berwick YB (288 miles) and later 4th Open Elgin (480 miles) etc. Nestmate of Uno Solo was Morning Glory, a silver chequer hen, that won 1st prize Thurso 1968 Essex & Suffolk Border Fed, and an Osman Memorial Trophy, as well as winning an East Midland Centre Gold Medal. Dam of 4059 does not appear, only I believe because it has been Bill Button's intention to project the evident importance he placed upon the silver line.

Sire of 4058 was 68R17694 a blue chequer cock known as Lane Again, a good 1st prize-winner for Bill Button that was bred from Bill Lane's champion Lerwick pair, 62293 and 62295. These latter two Lerwick performers achieved in the opinion of Bill Button two of the greatest performances from Lerwick in his time. Dam of 4058 was 41462 which was a daughter of 62233 and 82780. For the record Uno Solo and Morning Glory were bred from blue chequer cock 62248 when he was paired to dark chequer hen 22903. Dam of 62248 was 65151, a silver blue, whilst sire of 65151 was 2251 whose dam was also 943 the silver chequer pied hen that was bred by Bill Alty a noted Ipswich fancier. Note the frequency that 943 appears in the breeding of the Button pigeons.

Although Bill Button stresses the importance of the silver line and his homework research has been carried out with this in mind it does not require the brain of a genius to evaluate the breeding lines of Harry Kennett's successful Rome hen. Despite the limited ancestry I am able to extend she was undoubtedly bred for the job. Her owner was according to locals most certainly greatly fancied by her breeders! Although he did not know at the time of entry for Rome what he may know now after he has read this chapter of Month by Month. but I am sure he will be delighted to learn a great deal more about his silver blue hen. Most certainly more than he had dared to expect.

When Bill Button gave those two 1972 young birds to George Parry he told him to race them and if they turned out to be a cock and a hen to mate them together. However, before George could do this he died. Continuing Bill writes: "His wife apparently gave her late husband's birds to a club member whose name I believe was Kennett. I met this chap at an Old Comrades Show in London and learned that he was doing fairly well with birds bred out of those he received from Mrs Parry. He also stated that he was getting several silver blues and silver chequers. Since I brought my Squills year books and stud books with me when I came to Australia I have been able to trace the breeding of the two pigeons that I sent to George Parry hoping that you will find them of some interest and help also to see where the line of silvers come from. You will note that we were old Army Pigeon Service blokes. These included Bill Hammond, Billy Armour, Alex

"OLD K42"
Red Chequer Cock NURP43K42. Bred by A R Hancock, presented to F W S Hall as a squeaker. Won nine 1st prizes, 20 times in the first four. Flown Thurso 500 miles, plus several thousand miles racing and War service. Sire of many winners. Sired ten youngsters at 13 years of age. Above photo taken 2 December 1956. Note his legs, he suffered badly from arthritis in his later years. His first youngster 56FW8779 was 8th London Sect Berwick NRCC.

King, Jack Erratt, John Crosland, John Ambler, Jack Adams, Tommy Crotch grand bunch of lads. We used to get together and have a good time at the Old Comrades Show. I am just an old Pommie enjoying life out here —but how I really miss my old mates". Well as you can deduce for yourselves the above is a glorious example of being fortunate enough through the medium of The Racing Pigeon of bringing to light sufficient information to justify as well as uphold the need to keep all the records you possibly can about your pigeons and their breeding.

How I would love to be able to trace and extend fully the pedigrees of all the arrivals recorded from Rome. Especially the Scottish arrival reported by Wilson Noble of Fraserburgh, Scotland. The Went Bros', pigeon used in the production of the British record breaker will doubtless contain many of the best winning pigeons ever flown into North London including the Brutons, and will not prove difficult to compile. The new record breaking distance of 1,260 miles, 1,274 yards is tremendous, who would have imagined in their wildest dreams (I most certainly did not) that Rosie and Bruce's record of 1,144 miles from Barcelona would have been beaten in a race from Rome. A nice gesture I thought by Wilson Noble to name his winner Sir Lancelot after his late uncle Lancelot Kinnaird.

From correspondence plus a number of phone calls there are lots of fanciers with sickness in their lofts in the Greater London area. Equally too I learn from various parts of the country North, East, South and West that the Metropolis is not alone with regard to the many worries associated with our racing pigeons at the present time. For many the moult is still not over, but even so for many who separated their birds early, condition and health should be abundantly clear. However, this it seems is not so in many lofts. Loose droppings appear to be the main complaint. Colds too appear to abound in many lofts at the present time. Yet others complain of loss of wing movement. Others too report severe respiratory problems. A great number of pigeons have it seems been put down. Others are being treated with sundry remedies.

There are many conflicting opinions expressed regarding what may be the real cause simply because several of the diseases associated with racers as well as the fancy pigeons, including in some cases show pigeons, show a marked similarity of symptoms. This has always been so. It is not something new!

Not all illnesses among the infectious diseases affecting pigeons can be diagnosed without a microscopie examination by a qualified vet, or a trained person. Unfortunately at the present time very few trained people exist in the British Isles. Although there is no shortage of vets, the number interested in the problems associated with the racing pigeon are far less numerous than those who deal with animals. In

Belgium the position is different. In Britain the majority of vets are only interested in dogs and cats. We in this country are badly in need of a change of attitudes by vets towards the racing pigeon fraternity if we are to resolve many of the problems with which we are having to face up to so frequently today. Having well informed vets would make the payment of their fees less painful.

Most importantly it is imperative that the correct diagnosis be obtained. In this respect all too many fanciers believe they can make an absolute correct diagnostic decision. All too often this has been proven wrong. Even the medical profession has been known to give, or make a wrong decision with a member of the human race. Often this is due to the patient describing the symptoms wrongly! Many fanciers seek out a treatment for their pigeons without any qualified advice whatsoever. Today more than ever before the pigeon fanciers of the United Kingdom and Southern Ireland must be made aware that over the past 35 years a great change has taken place. Especially this is true regarding the many and varied remedies that are available and advertised repeatedly. Often through the power of clever advertising we can, if we are not on our guard, become brainwashed into believing that we must regularly make use of the various remedies advertised in order to prevent the diseases that are associated with the racing pigeon. The regularity and frequency of their usage could well be our greatest danger.

Today the facilities for the purchasing of veterinary remedies for the treatment of tricomoniasis (canker), coccidiosis (coccidia), salmonella and a number of other complaints, could be the downfall of many lofts. Only recently I learned of a fancier's experience after finding a veterinary surgeon who had taken an interest in his pigeons' problems. They showed evidence of "going light". After several visits and a number of dead pigeons for which post mortems (autopsies) were carried out, the fancier was advised to eliminate all his pigeons and make a fresh start. The reason for this, as given by the vet, was the pigeons had lost all their resistance to treatment owing to the damaging effect of too many regular but incorrect treatments of drugs and antibiotics over the past several years.

The great scourge at the present time appears to be that of watery droppings. In the majority of cases, I am sure it is mutated paramyxovirus. This is, in layman's terms, a form of rebellion against the vaccine (vaccination). The antibodies developed through the vaccine are being attacked. Unfortunately several who have resorted to a further vaccination have actually killed their pigeons. There is no time for the victims to build up other antibodies to counter the mutation. According to a learned friend, who wishes to remain anonymous, a further vaccination should not be considered when a mutated form of paramyxovirus has developed. The best treatment in his opinion is to treat these watery droppings (mutated paramyxovirus) as follows. For three days administer one level teaspoon (5ml) of aureomycin to one gallon of fresh water, followed by three days of glucose at the rate of two teaspoons (5ml) per litre of fresh water, followed by five days of metatone at the rate of one teaspoon (5ml) to two litres (three and a half pints) of clean water. All vessels should be thoroughly cleaned with boiling water before use. Glucose is quickly assimilated into the bloodstream. It assists the muscular system. Moreover glucose strengthens resistance to infections, as well as helping to cleanse the bloodstream. Pigeons that have not been given glucose before will not take readily to accepting "sweetened" water. You simply have to make certain they do take their glucose by creating a thirst. Withdraw their water and withhold for a period after their feed. It is imperative that you take the trouble to create a thirst. You have to be cruel to be kind. After the three-day glucose period, you follow this immediately with the metatone for a five-day period.

With regard to the veterinary profession, in all fairness fanciers too must do their part in bringing to the notice of vets that we do need their help in the establishment of a speedy service for the diagnosis of the various diseases associated with the racing pigeon. If this is not encouraged then it will no doubt bring about the establishment of a similar type of service that exists on the Continent, particularly in Belgium. There prominent wholesale corn manufacturers employ full-time vets whose sole task is to diagnose diseases in the racing pigeon. No doubt there are a few fanciers, even corn merchants, willing to establish a similar service in this country. There most certainly is a place for such a service. It will cost money, and time, and study, but for the right person it could prove both satisfying as well as profitable. What a fascinating subject it is. I only wish that I was 30 years younger.

There is without doubt great scope for further education in the study of pigeon diseases, especially canker. Trichomoniasis as it is known among students and qualified persons, and usually as canker by fanciers, is one of the most serious threats to pigeons. It can kill and it can also cripple any chances of a pigeon being able to race home. Canker is the most serious threat to form. Canker can impair the upper respiratory tract, and it can also seriously impair the muscles. Canker can also attack the navel. This is most noticeable in squabs and very young pigeons. There are a number of strains of canker and it is not always easy to diagnose it. You really need a good microscope as well as knowledge of your subject. Although canker can and does cause serious losses among turkeys, chickens, ducks and even canaries, the pigeon is the most common host. However, in each of the various varieties of birdlife mentioned they are in turn more than likely infected by a particular type or strain of canker. The study of canker is probably one of the most inexhaustive and I would consider it to be a most fascinating subject. Harkanka, that used to be marketed by Harkers, was one of the most useful forms of treatment against the disease. It is a great pity that it ceased to be manufactured. In capsule form it provided a most effective individual form of treatment. The loss of this product proved a considerable disservice to the sport.

The health of our pigeons is of the utmost importance. It is essential to success. We therefore must at all times be fully prepared to obtain the best advice we can. Therefore, it is important that we find out who among local veterinarians is interested in pigeon diseases. On the whole pigeons are hardy creatures, yet there are times when one cannot be certain what a pigeon is suffering from. A pigeon suddenly dies for no apparent reason. An autopsy would tell us a great deal. If you take the trouble to contact your local MAFF office they will either help you or give you the information you require as to the best way of dealing with such matters. There are so many diseases that have similar symptoms. All too often in recent years many of us have through circumstances, or lack of knowledge of the facilities available at Ministry level, become self-styled vets. Often our limited knowledge, or lack of observation, or failure to interpret what we have observed, has proved costly in terms of infected pigeons. As I have written before the racing pigeon is very hardy. However, you must always be on your guard, ready and willing to seek help when it is needed most.

If you are uncertain of the disease, or sickness that is troubling your birds, do not ever be afraid to call upon your local MAFF office — it is there to help you. You can call upon your local vet or make an appointment. But be organised. Take along a sick pigeon. If you should be unfortunate to have a dead one, take it along discreetly wrapped. At the same time take along a healthy looking live one as well. It would also help if you took along with you a sample of droppings taken from within your loft, you only need a small amount. Either use a small screw top jar, or a well made clear polythene type see through bag. It will cost a reasonable sum of money but it will prove worthwhile. Remember you often require a microscope in resolving the identity of the various kinds of protozoa that cause disease. The veterinarian has such appliances.

What is more certain he knows what to look for or can identify what is seen. Moreover, his or her trained mind will help you to resolve your problems. As I repeat once again pigeons are on the whole very hardy. They are not a menace to public health so you need not be afraid to call upon the vet to help you when you need it most. Recently some of the commercial firms have started advertising help lines, try them. You can also obtain a good book on the subject of pigeon diseases. Healthy Pigeons by Dr L Schrag is most excellent. So too is Keep Your

Pigeons Flying by the late Dr Leon F Whitney. I hope it will be reprinted soon. There is a great deal more that can be written by the layman for the layman but this will have to wait for another effort on my part. However, if those who have a mind to help the Fancy there was never a more appropriate time to invest in a good microscope.

Remember healthy pigeons are happy pigeons. Successful fanciers are those who see to it that all their pigeons, are both healthy and happy. That is the reason for their success. If you are worried about your pigeons' well-being, or the lack of it, do not be afraid to seek help. It is not a crime, or a sin, but it could be both if you do not do something about such problems. In my efforts to pinpoint the problems that fanciers of today are faced with I hope no one will take offence over what I have written. It is composed with the best of intentions and my great love for "creatures great and small".

OCTOBER

The Moult — Separating — Alf Baker Pure strains — Stichelbauts — Huyskens-Van Riel — Pigeon Management

The weather we are enjoying has helped in the steady yet seemingly never ending shedding of the feathers, which are falling as freely as are leaves from the many splendid trees that still abound thankfully. A good moult especially for racing pigeons is still very important, good nourishing clean food equally so. Many times have I stressed the importance of the moult, yet it was not all that long ago, maybe 60 or 70 years, when many fanciers considered the moult was a disease! Yes they really did. Yet the remarkableness of the moult is not confined to our racing pigeons. All birds are affected by this phenomenon.

It also is a natural order of events in the majority of creatures, at least so far as animals and reptiles are concerned. You can without fear of being wrong more or less consider that trees too moult for that is happening now as I type these lines — their leaves are falling and whenever the slightest breeze arises leaves are to be found in plenty.

All animals regularly shed their coats (hair). Snakes slough (cast) their skins (scales), scales being the forerunners of feathers developed by the creatures that emerged from the oceans. Those that remained in the oceans retained their scales. Remember too that both leaves, as well as, feathers, can be utilised in the cultivation of the soil. This is nature's recycling process. Their replacement, where applicable, denotes a renewal of life and consequently their renewal demands plenty of good nourishing food. The quality of their replacement is dependent upon the quality of the food available, this is why the moult is helped enormously with a rich variety of sound clean corn and oil-bound seeds, this is where linseed and safflower are so invaluable. The real genuine stockman leaves no stone unturned in order to bring about a successful conclusion to the moult.

The bath is very important, the sun too plays an important part in the production of good feather. Meaning that facilities for bathing and equally sunparlours, an aviary or open topped pens for sunbathing for those who keep prisoner stock is essential. Stock that is flying out can be dealt with more simply and conveniently. As I have stated already no privation in the matter of good quality food should be one's policy during the period of the moult. This state of affairs brings us to the question of separation of the sexes. If you separate then you indeed have no need to restrict variety and within reason the daily amount per bird of corn. But remember this does not imply that you must stuff your birds constantly. The advantages of separation are therefore twofold, yet there are many, including the writer, who has practised non-separation without breeding and egg laying problems, but this does require removal of all nest boxes, or where fixed boxes are incorporated firmly closed. All feathers should also be removed as frequently as time will permit, at least once or twice a week or daily where possible. Whenever a pair are noticeably ever ready to canoodle then be prepared to place a brick, or anything that prevents this state of affairs developing. There are without doubt advantages either way. There also are distinct disadvantages. If I had the facility of removing all my nest boxes I would do so. The extra air space the removal would provide is one of the advantages. More convenient too, is the facility such removals provide for you to renovate the boxes, both within as well as without. Another drawback with non-separation is that when the weather is very mild one really does have to take care about the feed. Overfeeding will undoubtedly encourage mating.

In Belgium many lofts place their hens in housing that is out of reach and earshot of the cocks. In Belgium buildings that do not incorporate perches of the box type are very popular. In some that I have visited particularly in Belgium, they use vee-perches, or saddle perches, whilst a number use only poles fitted across from wall to wall. These latter I consider are by far the best type for the preventing of hens mating together. Such sparsely equipped places have much to recommend them where large numbers of hens are maintained. Dry and warm, yet airy, but lacking any type of equipment that may lend itself ideally for two hens to canoodle, and subsequently lay. Once a pair lays it is not too long before others mate and follow suit, sitting in turns, the four eggs that follow. This is also the inconvenience that deep litter brings about, and the reason that not many of the Continentals practice a deep litter policy. Once hens do take up mates they will forever after continue the practice. Even cocks will, to a

OCTOBER

certain degree mate between themselves, particularly if you deny them hens for any length of time.

However, there is not the slightest doubt at all that where a complete separation of the sexes is steadfastly applied the work is doubled. Feeding the birds together, exercising together, bathing together and cleaning out all at the same time reduces the workload, you only have one loft to maintain. There is, however, no doubt whatsoever, as I have noted over the years, a far happier state of affairs in the loft where non-separation is practised.

Doubtless for the obvious problems I have pinpointed you cannot neglect the need to be vigilant, especially when we experience the type of weather we have done since Global Warming became a problem for mankind, as it undoubtedly now is. Of this I am truly convinced that so far as the racing pigeon is concerned non-separation of the sexes creates a far happier home life and for this I know from past experience the moult too benefits. Pigeons that are kept together are far happier than those that are completely and utterly removed from the sight or hearing of each other. The time to separate will be a fortnight before your mating update. A word of warning, remember that when your birds are separated you take steps to ensure that in the case of those that you have purchased from another loft and broken out to your own loft these birds are not allowed to be released during the period of separation. Hens particularly, given the opportunity, will return to their old home whenever the pain of separation is thrust upon them.

Similarly cocks too, will also fly back to their former home whenever the opportunity presents itself. In this day and age of keen competition it is in my opinion not wise to practice a policy of an all barley diet once racing is over. Unfortunately far too many do. All too often many of these fanciers find it difficult to maintain a winning loft when the longer races come round, if you really must then there are several good depurative mixtures available today. At this time of the year a good mixture containing good wheat, a little sound barley, linseed, safflower, beans, peas, even a variety of peas, also maize, dari, milo, sunflower and a small amount of condition seed, plus a few pellets, will help your pigeons through the period of the moult.

Once the moult is completed then and only then will it be necessary to reduce and change your feeding, but not before. At this time of the year the number of meals per day depends upon your workaday life. It will soon only be possible to feed your pigeons once a day. During the longer periods of confinement that follow immediately after we put the clock back make absolutely certain that grit is before them.Fresh vitamins can be made readily available if you remember to provide your pigeons with at least a weekly ration of fresh greenfood, or in its absence a supply of sprouting peas, or tic beans, or even both. Sprouting corn is one of the finest forms of fresh vitamins that you can possibly give to your racers.

Frequent baths too are imperative. Baths really do invigorate the system and most certainly pay dividends at the present time — facilitating the expulsion of the old feathers is a positive must. No matter how severe the weather may become, pigeons still enjoy the pleasures of bathing. The successful fanciers are those who carry out a strict regular routine throughout the late autumn and winter months. When winter is upon us and calendar-wise the shortest day marks its beginning (December 21) then is the time to consider a reduction in the amount of contents of food per bird per day. Especially so if the winter should prove at first to be very mild.

Again the method or routine for mealtimes that you follow is largely dependent upon your own workaday life. Often in the case of partnerships you can the more easily make suitably convenient arrangements. It is possible that one meal only per day is sufficient, midday is probably appreciated by the pigeons. On the other hand you may prefer to provide two meals per day one in the morning, the other in the evening time provided you have lighting facilities.

My own preference is for a third-part feed early in the day, and two-thirds balance for the later meal. Many successfully feed once a day only — midday. If pigeons are allowed the use of cafeteria like hoppers they will make their own regular mealtime. One fancier who springs to mind, and whom I have known for a long time now, feeds his pigeons a full crop once in the morning and only the very smallest titbit late afternoon. He is without any doubt at all a highly successful fancier, and always the one they all have to beat! His name is Bob Taylor of Southgate, and when I remind you that the Southgate club members included the likes of Bob Hutton and Pete Pedder, as well as several other successful members who annually collected at the Southgate Social prize distributions! Bob Taylor's system was described for all to read and study in the 1989 Squills Year Book, written by The Red Baron.This particular article entitled "Chalk and Cheese" describes the systems of the two most formidable competitors that the rest of the members of the Southgate Social FC have to beat. It is well worth reading over and over again. A contrast in system, methods, and personality that is almost unbelievable. However, the one feature of this article that stands out is this, basically, both fanciers are fundamentally one ounce a day fanciers.

To complete my résumé for autumn and winter the question of separating for next season's matings is always a debatable matter when fanciers get together. Unfortunately this is not a feature of club meetings which would I am certain prove a boon to novices and the lesser successful. It is always important that where fanciers do keep their pigeons together up until the third or fourth week of December,

OCTOBER

or as some do New Year's Day, that you make a point of complete separation, for at least two whole weeks before mating time. For those who prefer to mate on Valentine's Day, or any other particular day, much of which is largely based upon your ambitions for racing, it is important that your hens are clear of egg laying possibilities, so that parentage can be considered a "certainty".

When pigeons are separated I know from experience they fly better on suitable late autumnal, and winter days. Some do not believe in exercising pigeons during the winter, and unless the weather is really ideal, they are wise to refrain, but in good days I like to have my own pigeons out. Often for many and for me too, I find it is much safer to fly the sexes on alternative days, hens one day, cocks the next. For safety reasons, when the weather is good I like to get my hens out in the morning, then if, as they will when at their fittest, they cause you any anxiety in a long absence you can let out the cocks. This is a safety measure that I like to have in reserve. So hens in the morning and cocks in the early afternoon. Better still hens one day, cocks the next.

At this time of the year there are many, now that all thoughts of breeding and racing are past, who turn their thoughts to the matter of next year's breeding plans. The line breeding minority know what they are going to do. Those fanciers who possess a successful family will continue to cultivate by line breeding to their successful breeders and racers. There is no better means of continuing success like breeding from successful lines. Try to retain at all costs a family that is already successful. This is gold medal material no matter how unfashionable they may appear to some!

Unfortunately all too often commercially fashionable strains through clever advertising often sway the less successful to bring in a completely different strain. The newcomers, and those who have not yet mastered the art of conditioning, training and general management of a loft of racing pigeons are easily influenced. All too many fanciers keep changing their minds about strains. Lack of success is often owing to lack of experience. All too often it indicates a thorough ignorance entirely. Not, as so many believe, the need for a different strain, or breed, but the need to be better informed. It does seem that I am forever beating the same drum. But I am more than a little anxious about the future of this splendid hobby of pigeon racing. Furthermore, in the world we are now living in, we old 'uns should do all we can to encourage the young people of today to take up a hobby. Naturally our foremost duty (because we know the fascination of pigeon keeping) will be to help swell our own dangerously low numbers. This is a serious problem that we should all try our utmost to improve. Novices and the lesser experienced need helping in this respect.

Yet first of all, you have to learn the basics of pigeon keeping. That is precisely what I meant when I suggested to be better informed. For the majority economics have to be considered in a sensible manner. The pigeon house, loft, or pen, whatever it may be called depending upon where you live, does not have to be elaborate. It need only be simple, yet soundly built, preferably portable, and built in accordance with local council regulations. It is not a bit of use putting up even a portable building unless you first of all apply for permission, otherwise you will run into trouble. Neighbours have to be considered.

Once these normally simple requests have been carried out and accepted officially your first and only major problem has been overcome. The loft does not have to be palatial. Yet it can have a smart appearance, as is illustrated in that splendid monthly magazine Racing Pigeon Pictorial. A new strain will not compensate for lack of clean water daily, clean food placed in clean and covered corn hoppers. Very elementary, but still important. A pigeon loft must at all times be dry so the roof is of the utmost importance. Importing a new strain because of lack of success is no use unless you have learned sufficient about the basic principles of keeping racing pigeons. One of England's truly outstanding fanciers, Alf Baker of Wood Green, built his very first pigeon loft in 1926 when he was 12 years old. Here in my opinion is a splendid example of good fanciership from a very early age — at its best.

Alf built his first pigeon loft with the timber he obtained from glass cases that contained Belgian manufactured glass. These cases cost the sum of eighteen pence each (old money) from the glass shop at the end of the road where Alf lived. It was at this very shop that young Alfred learned his trade as a leaded light maker and glazier. By the time Alf Baker was 14 he had been studying the pigeons he had got together every day without missing a single day of observation (for two whole years!). Without any doubt whatsoever young Alf was a born stockman. As soon as he left school Alf joined the local Sunday morning club that was not affiliated to the then ruling body, National Homing Union as the RPRA was then known. Sunday racing was not allowed by the NHU unless a holdover happened. This club had a half-mile radius. In those days Alf has since told me that nearly every house around about kept pigeons or was friendly with those who did.

The race fee was 3d (1½) per bird. The shorter races were convoyed by a local lorry owner. When the races got beyond 100 miles the birds were sent overnight by passenger train. The longest race was Perth, over 350 miles to young Alf's loft. The very first great event at this distance and probably Alf's greatest thrill, despite the many outstanding successes since (including winning of the great London NR Combine five times), was in winning that Sunday morning club's Perth race in 1928 with the only bird on the day. The members could not afford clocks, although they were in use in clubs affiliated to NHU, so competitors had to catch and run to where a club official was waiting to verify the birds by their ring numbers entered into the club's official ring book before the race.

The 1928 Perth winner was a red hen. The next year the Red Hen as she had been named was again the only bird on the day from Perth. Some pigeon! No wonder Alf has always liked racing hens and still loves a good red, be it cock or hen. The 1929 Perth race was a real teaser, the Red Hen dropped at 9.15pm at night in pouring rain. Alf once wrote she was so wet she looked like a dark chequer! The parents of this dual Perth winning hen were bought by Alf at a Sunday morning auction sale conducted in one of the Sunday morning club's members' back garden, for half-a-crown the pair! After this success with the Red Hen the man from whom young Alf bought the "Half-a-Crown" pair was ready to purchase them back for a huge profit, but young Alfred Baker was unmoved.

It was soon after the second Perth race win that Alf's father who had a good friend named Ginger Taylor, who was also a very good pigeon man, between them, acting upon young Alf's advice, composed a letter and wrote to the famous Belgian fancier Mons Stassart who at that time was probably the best fancier in Belgium. The aim was to purchase half a dozen Stassart young birds. The pictures that young Alf had seen so impressed him for type and size that the Stassarts were his choice for a foundation strain and no other would satisfy him. Eventually six youngsters were obtained, two of these were red hens, two blue cocks, a chequer cock and a chequer hen. All six had to remain prisoners, they cost too much to risk being allowed their liberty especially as they carried Belgian rings. Young Alf Baker was more than satisfied with his Belgian imports and fully prepared to establish his family of pigeons on the strain he had chosen. Here was patience, determination, ambition, stockmanship all in one and at its best.

About 1935 or 1936 Alf joined Wood Green HS as a novice. Although he had won a number of races in the Sunday morning club these had not been flown under NHU rules and were therefore not recognised. Alf used to look forward to the meetings and never missed one. It was exciting for him to be able to listen to the older fanciers. The president was a fancier known as Spratty Drayton. Harry was really his name, but he was a good pigeonman as well as an expert on wild birds too, which Alfie was also keenly interested in. Spratty was the proud owner of a flight of redpolls that he used to let out for a fly and they would return to their aviary. Alf became very friendly indeed with Mr Drayton and was privileged to see and handle his pigeons, as well as witness the "flight of the redpolls".

As long as I have known Alf Baker he has always maintained how much he learned about pigeons from his friend Spratty Drayton, who

London Social Circle's 1948 Thames river trip, secretary Frank Hall, is at the front in braces with president Harry Smith, also in braces, far right.

in Alf's own words was always a hard man to beat with pigeons. Mr Drayton was the only man Alf had known who had one of his pigeons set up by a taxidermist and placed in a glass case on his sideboard. Mr Drayton always described her as one of his very best pigeons ever. She had won the Banff race three years in succession. Whenever he spoke about his Banff Hen his eyes filled with tears, that was the kind of man he was, and in Alf's own words the reason he was so hard to beat in the races. When Mr Drayton died it was a sad day for Wood Green club as he really was a great president. In the words of Alf, "Harry Drayton was a great leader. And I was greatly honoured when the members asked me to take his place as their president."

The six Stassart youngsters included at least three grandchildren of Mons Stassart's famous Baladin, as well as a red hen daughter of Baladin. She was the dam of Plum, the mealy cock, probably the greatest of all the Baker pigeons. This famous pigeon (Baladin) was never a good handler, yet he was a formidable racer. Alf recalled the alarm that was created when Baladin went missing for five days, after he had been retired from racing. The loft was built over Mons Stassart's restaurant and Stassart was convinced Baladin had gone down one of the chimneys. All the bricks were taken away from the fireplaces and on the fifth fireplace, once the bricks were removed, there was the famous old pigeon none the worse for his five-day imprisonment without food or water. So proud was young Alfred with his six Stassart youngsters they were never ever let out and were breeding outstanding stock up to and including eight and nine years of age.In fact, with the youngsters bred from these Stassarts in 1935 Alf joined the Wood Green HS. Those youngsters became the progenitors of many wonderful pigeons that Alf Baker produced in the years that followed. From 1947 to and including 1952 Alf won well over hundred 1st prizes from 50 miles to 500 miles.

Among the famous pigeons produced was Plum, a mealy cock number 158, probably the greatest of them all, and a winner of seventeen 1st prizes, also winner of the London NR Fed five times. Plum sired at least the winners of twenty-five 1st prizes. Of the seventeen 1st prizes Plum won, eight were won in successive races. Plum never carried any flesh at all. In the last race Plum won of the eight successive wins, from Berwick, the day of the race opened up fine, birds were up at 7.30am, wind SW. In Alf's own words: "I knew it was going to be a hard fly. At 10am it came over black, started to rain and never left off all day. I was sitting down cosy-like having tea when my daughter said, 'Plum, dad'. There he was, shaking the rain off. I clocked him at 5.32, result 1st Club, 2nd London NR Fed. He really was a great pigeon."

As a result of the many hours Alf would spend studying his pigeons he was always able to know the temperament and characteristics of his pigeons. Especially the pigeons that showed intelligence. In the case of Plum Alf Baker noted that he was always after his own mother. She actually went barren the year she bred Plum, so thanks to his keen observation at all times within the loft Plum was officially mated to his mother and sent him calling to nest in the first race, after winning the race, upon his return from the clubhouse Alf would put a pair of warm eggs under Plum's mate (his mother) and Plum used to race home and go straight on the eggs till the eggs were taken away on a Wednesday morning to get him calling to nest again. He won LNR Fed five times this way, indeed all his races were won in this condition. That is observation used at its best, dedication to the task of winning, the hallmark of a good fancier.

Over the years Alf Baker noted that if a cock raced well that way so did the majority of his sons and grandsons. All this has been written for a purpose to let the youth of today realise that success in pigeon racing is down to hard work, a keen observation, a willingness to do all one can, at all times, regardless of other interests. Fancy lofts, Sputnik traps, and all the many contraptions available will not take the place of good fanciership. You really have to get down to the real tasks of preparing your pigeons with complete thoroughness of purpose.

Alfred Baker of Wood Green, after the 1939-45 war only had one eye yet despite this handicap he did not miss much when he was in and out of the loft studying his beloved pigeons. A Thomas Long Trophy winner, in addition to winning the Combine five times, Alf was also 2nd Open Combine five times, also Osman Memorial Trophy winner in LNR Combine Thurso race, NHU Gold Medallist, and winner of coveted Championship of London Rose Bowl. Because I have known Alf Baker so long, and fully aware of his ability as a pigeon fancier first class, I am glad to hear that Colin Osman has persuaded Alf to write that book he has from time to time talked about. It's the kind of reading that would be of the greatest help to novices, the unsuccessful, and even those who have a reputation themselves. Since writing this the book "Winning Naturally" by Alf has been published and what a wonderful read it is.

Without a doubt Plum was a marvellous pigeon. Many of the latter day champions come down from Plum. Such pigeons as 1724 winner of many prizes including at least nine 1st prizes, different races, 3rd Open LNR Combine. The blue cock 1724 was a son of Plum, who was a son of one of the original six Stassart youngsters, the Red Hen daughter of Baladin.An interesting story emerged the last time I spoke to Alf. On the day the Germans were doing their utmost to obliterate the London docks, a full scale air-raid in full swing Alf was waiting for Plum to get home from a Northallerton race, one of those eight races Plum won in succession. Alf's next door neighbour called out to Alf, "You ought to be in the shelter there's an air-raid on." Alf's reply was, "I'm waiting for Plum and I am not going to stop looking until Plum's safely home." Home he came too, winning the race, and very well up in the Fed too.

After considerable thought I have decided it will help fanciers especially the young and those who find it difficult to evaluate the written word if I summarise.

No 1. Observation is of the greatest importance, and if like Alf Baker, you are a natural born fancier, you will appreciate the value of continuous daily observations, upon your pigeons, year after year, that helped young Alfred Baker to improve his skills enormously.

No 2. Concentrate at first upon one strain, a successful one, as the Stassarts undoubtedly helped Alf Baker enormously. Too many different strains of all sorts is a grave mistake. Any cross deemed necessary as Alf Baker has proved beyond doubt are light years away! Especially as Alf had discovered the prepotency of Champion Plum, although perhaps at the time without being fully aware of the phenomenon or prepotency! What he had discovered was Plum's great ability in passing on the genes of racing ability, meaning, in fact, to be superiorly powerful of prepotent. After all Champion Plum, bred in 1940, was a grandson of Mons Stassart's great Champion Baladin, through the imported Red Hen that never flew out in her life and was kept in the attic. This daughter of Baladin was a goldmine. This imprisonment should make some of those impatient fanciers, think hard, when they think about the outstanding sons and/or daughters they have lost through taking silly risks with expensive stock that were too old when they purchased them as young birds.

No 3. Alf Baker, whatever the time, day or night, tried hard never to overlook the smallest detail in order to secure victory no matter what the status of the next race may be, the pigeons he intended to send had to be at their best. For instance, how many fanciers ever think of the possible consequences of flea powdering your pigeons when sending them to a race? This is a grave mistake, yet I know many who do it to this day yet never consider the consequences of the inhalation of chemical powder en route to the race point.

No 4. Alf Baker believed in the education of his young birds, yet he would not despise a well bred intelligent late bred. For example, I recall more than 20 years ago Alf Baker describing how disappointed he was one year when rearing a few special late breds from the Stassart bred imports.

He put down five of the 12 he reared without any hesitation whatsoever. Late breds have to be both physically and constitutionally sound. Inbreeding demands this ruthless elimination. There is nothing wrong with inbreeding so long as you are ruthless in discarding those found wanting, and Alfred Baker recognised the importance of such a policy at all times. And when, as I often do, recall that Alf Baker, the wizard of Wood Green came nearer to winning the historic London North Road Combine more times than any member in its long and glorious history he must be the man you should take notice of, for the living legend almost won it ten times! If you want to read more about Alf, and you ought to, you can read it in his autobiography "Winning Naturally" published in 1991.

A fancier enquired as to "Which Belgian strains can I advise him are the pure strains?", adding, "or should I look to France, or Holland?". The answer is simple for there really are no pure strains in any of those countries. On the other hand there are several very good line bred to champion breeder strains, as well as line bred to champion breeder/champion racer strains. Actual pure strains where one would expect to find that every single individual looks alike, and from the written or spoken evidence, are without a single introductory cross for at least 20 years are as plentiful as a snowman in the desert. To achieve such a phenomenon would imply as in nature that every single individual member of the group were related as well as being identical in type and colour, so far as the racing pigeon fraternity is concerned purity of strain does not really exist in this way. Any fanciers who claim that they really do possess a pure strain would be very hard pressed to substantiate such claims. Whether it be in Belgium, France, Holland, Germany, the United Kingdom or for that matter anywhere else in the entire universe where pigeon breeding for racing is practised.

Another fancier enquired from the regular study and constant scrutiny of the articles and advertisements contained in The Racing Pigeon and Racing Pigeon Pictorial it would seem that the most pure strain must be the Stichelbaut, whilst my friend who has kept pigeons for some time now contends that the Dordins are the purest strain. My answer to both is neither of these strains were pure and neither the late Alois Stichelbaut or the late Pierre Dordin, to my knowledge, ever claimed that their respective families were pure. At least if they ever did .I have never personally read any claims to this effect. Nonetheless it would probably be fair to state that the painstaking work carried out over the years, plus the cleverness of both fanciers, coupled with their continuously applied selective breeding policy established without doubt certain qualities and characteristics that enabled fanciers to recognise each family group or strain whichever is the accepted terminology.

These qualities are still recognisable wherever these two strains are maintained. In fact, in certain hands wherein line breeding to quality reproducers within the group or strain is maintained, and there are still a few who practise this form of cultivation, you can immediately recognise the characteristics established by the founders of each strain whenever these emerge as they must when line breeding is carried out. If you also make a study of the breeding details you will learn to appreciate sooner rather than later that both Stichelbaut, and especially Dordin line bred as much as possible to their best breeders, and even the experienced fancier with a firmly established family knows only too well these are not too plentiful. Both Alois Stichelbaut the Belgian, and Pierre Dordin the Frenchman, having discovered their best breeding lines that had descended from their basic pigeons never neglected to recognise the value of such key pigeons. There is no doubt that both fanciers set themselves high standards.

This was certainly true of Pierre Dordin who was always working and planning for the improvement of his family of pigeons by raising the quality of his birds. For Dordin the raising of the standard was of paramount importance. For those who are acquainted with either family or strain group, or as some do cultivate both, will I feel certain be fully aware that it is certainly not difficult to be able to recognise the two strains or groupings whichever you keep. There is a marked contrast of type and looks in these two very exciting and indeed very highly successful family strains. On the one hand the Stichelbauts are mainly black velvets, black chequers, dark chequers and blue chequers with the odd pencil blue and in each the odd white flight. A glorious example of being the world famous Ware Ijzeren (Red Iron), that was a black chequer white-flighted cock Belg 57.3064724 winner of 1st National Libourne, 2nd National Montauban, 3rd National Bordeaux, 11th National Dax, 1st Ace Pigeon of Belgium. A truly wonderful pigeon because not only a first class racer but also a remarkable successful breeder and you do not come by these very easily. Have also noted the Ware Ijzeren described as Old Iron as well as True Iron but each with the correct ring number.

Further to the above colour observations you will frequently find the Stichelbauts are bronze tinted, particularly in the bars, and these too are often distinctly widely barred. It is a characteristic of the strain. They are the colours of the Stichelbauts that have descended from those many of which were similarly marked among those 40 or so pigeons that so far as I can research comprised the entire one and only public auction of the Alois Stichelbaut family which actually took place on Sunday 27 January 1946. There were no reds, no mealys, no grizzles, no self-blacks and very few blues. In the genuinely concentrated Stichelbaut pigeons that exist today in the UK the majority are also black chequers, dark chequers, blue chequers and the odd slightly pencilled blue. As well as these are the odd blue with tinted bronze in the wing bars, and equally, too, the odd white-flighted pigeon, and a number too that carry that characteristic bronze tinting, and also the wide bars in the chequers.

In fact, it is probable that today there are truly more purely bred

Stichelbauts in the UK than there are in Belgium. But only pure through a more in-depth concentration of Stichelbaut lines. Another outstanding Stichelbaut that readily springs to mind is the famous Remi of 54, one time the property of Descamps-Van Hasten and later, when fairly old, purchased for a considerable sum by Emiel Denys. What an investment that turned out to be for Emiel. Again the success of Remi of 54 showed only too well the value of preserving the breeder lines within a family, something that I have advocated for years and years rather than continue to make a wide variety of new purchases of completely unrelated pigeons as so many seem to do. Many of the best lofts in Belgium value the Stichelbauts as a cross. For example, G & M Vanhee and the late Pol Bostyn all valued highly the Stichelbaut as a cross. But I assure all and sundry that I doubt if any pure strains exist in Belgium today or for that matter ever have, even the Gurnays and the Wegges contained crosses.

Several years ago I obtained a photocopy of the Stichelbaut auction sale announcement through Jim Biss after he had made several visits and purchases from Descamps-Van Hasten, coupled with the help of Van Hasten's loft manager Camiel Donkels, it read as follows: "On Sunday 27 January 1946. Compelled important auction of all pigeons (apart from two couples to keep the breed) from Mr Alois Stichelbaut of Lauwe. As written in our previous edition, Mr Stichelbaut is at present unable to take care any longer of his pigeons. Furthermore he cannot rely anymore upon the help of his wife, who has four children. Therefore, it is with much regret that he has made this decision. Considering that Alois cannot live without pigeons he decided to keep two couples of this wonderful breed". I remind readers that this announcement was not written by Mr Stichelbaut but by the person responsible for the publication of this announcement for the sale. The sale announcement goes on, "The latter are not the great prize-winners, to the contrary, these four pigeons were bred during the war, two of these were bred during the years of 1942/43 and two in 1944. In 1922 the origin of these versatile race pigeons were founded from birds bred by the late champion Mr Alf Derumeaux who was a good friend of Alois, and others descending from the Bordeaux specialist Mr Camiel Christiaens from Stacegem, the late Camiel being an uncle of Alois. In 1927 new blood originating from Vincent Marien in Antwerp was introduced. In 1930 the late A Vandecandelaere pigeons were bought and these were of the breed of Lagae-Blondeel. In 1935 a breeding of the famous Bordeaux flier Armand DeClercq of Lauwe was introduced.

The above are the fundamentals (foundation) of the pedigrees of the famous fliers from Mr Stichelbaut". The books compiled by Jules Gallez contain more including a list of the basic pigeons of the Stichelbaut family. Further information showing the value placed upon the Stichelbauts is given in the succeeding volumes of the History of the Belgian Strains. The photographs too will help you who are Stichelbaut enthusiasts by presenting a splendid impression of the rare coloured dark chequers and the blue chequers. Especially study the photographs contained in History of Belgian Strains — Part 1.From the crossings used by Alois Stichelbaut he produced a remarkable family that are even to this day fully appreciated by the Continental fanciers particularly the Belgians of whom many are known to have cultivated in part, or through a single introduction of the Stichelbauts, which adds further credence to the statement of years gone by that top photographer Anthony Bolton has also since repeated to me, "that few top Belgian fanciers do not include the Stichelbaut in their introductions whenever they consider a cross is required".

As a contrast to the characteristic colours and average size of the Stichelbauts, the Dordins were of large handsome type that comprised a majority of glorious powder blues with a sprinkling of dark chequers, blue chequers, mealys and the odd red chequer, as well as a few white-flighted pigeons. The beauty of the Dordins for those who do not keep the strain can be best appreciated by getting hold of a copy of Dr Tim Lovel's book, namely Pierre Dordin — The Complete Fancier, His Life, His Pigeons, His Studies. The foundation of the Dordins was based on the early Hansennes, the Commines strain, the Grooters, as well as pigeons from the loft of Felix Rey of Anderlecht, whose family base included the Grooters and the Wegges. Here again as with the Stichelbaut, Pierre Dordin valued highly the breeders and always made a point of mating his best racers to the children of his best breeders. The beauty of the Dordins has to be seen to be fully appreciated. Even the big handsome ones win well in the right hands.

Hopefully the above has helped to answer those who have sought advice about purity of strain and the Stichelbauts and Dordins in particular. There are many other old established families in Belgium notably the Cattrysse, the Delbars, the Van Bruaenes but they are not pure. But remember there are still several very fine line bred families in the United Kingdom that are still very hard to beat, especially when it comes to long distance racing. Take for example the Logans, the Savage Barkers, those great pigeons, the old Brutons. Another family too is the old Westcotts, which are still going strong. Nor must we overlook the old N Barkers, and that splendid old strain the Jarvis, founded by the famous Frank Jarvis of Harpenden. Then again what about those great favourites still, the famous Kirkpatricks, they too are still hard to beat.

Without any doubt all those United Kingdom families or strains I have mentioned above together with the Barkers are line bred by those who specialise in the old English strains. Neither overlook the famous Westcotts, still among the most successful. None are pure in the strictest scientific sense yet they still possess characteristics that are well known and recognisable by those who specialise. Old fashioned they may be — but outdated never!

Recently I received a phone call from a fancier enquiring if I had ever heard of a strain known as the Cools! At first I thought I could help him by remembering my much-thumbed copy of Production of a Strain written by the late Lt. Col A H Osman somewhere around 1924/25, but alas it was not mentioned. My old equally much-thumbed office copy of the Lea Rayner book Creation of a Strain mentions the Cools' strain but only very briefly. However, I am certain the strain was well known before the 1914-18 war and that M Cools came from Brussels. In those years strains like the Jurions, Grooters, Soffles, Bovyns and Cools were popular, and, of course, the N Barkers. These were all Belgian strains.

Cools are probably the least known of the Belgian strains I have mentioned so far as UK fanciers are concerned. One man who I have read about many times was R Gough of Bow, East London, from whom the famous F W Marriott bought a wonderful pair of stock birds that figured prominently in Marriott's early champions. Mr Gough most certainly used the Cools' lines. I know I have read this somewhere but cannot for the life of me remember in what book, or publication it was published. However, I do recall that although the Cools' strain was well known and esteemed in Belgium the strain only came to this country in small numbers, and one of the few UK enthusiasts was the highly successful East Londoner Mr Gough who sold pigeons of the strain of Cools to a Reading fancier whom I have been unable to trace. It was from the Reading fancier that Sir Oswald Mosley, Bart, purchased descendants of Cools imports. Sir Oswald lived at Rolleston Hall, Burton-on-Trent.All those Belgian strains mentioned earlier were kept at Rolleston Hall, as well as some of the English strains, including the Griffiths, Stanhopes, Taylor Gits, and most certainly J W Logans, Tofts and the John Wright of Great Budworth strain which was influenced considerably by the early Jurions. Often I wonder how many fanciers even of the old school can remember Sir Oswald Mosley, leader of the Black Shirt British Fascists, as an advertising pigeon fancier, who employed a loft manager by the name of Niddy, or Niddry?

The Huyskens-Van Riel

Another fancier enquired about the Huyskens-Van Riel strain and where he could obtain them. They do exist in the UK but only in isolated pockets. My old friend the late Harry Ashman, father of Bob Ashman the 1987 Midland National FC Angouleme winner was always a great enthusiast for the Huyskens-Van Riel strain so too was the late Jack Humphreys of Tottenham, who had a visit several times from the famous Dr Leon Whitney who too was a great admirer of Huyskens-Van Riels. Bob Ashman too, knew the value of the Huyskens-Van Riels. Geoff Kirkland too has used the strain with considerable success. So too has Billy Maloy now back in Tottenham from Norfolk where he moved to soon after he won the London NR Combine from Berwick. Lennie Brooks, a one time Londoner settled for several years now at Spalding, Lincs, is another who cultivates only the Huyskens-Van Riel lines, and when he has a mind too, races quite successfully.

This was the strain that dominated the Belgian scene in Antwerp for a number of years after the 1939-45 war. As one fancier seeking information on the Huyskens-Van Riels observed recently: "I have never seen anywhere anything concerning the bloodlines used in the foundation of the Huyskens-Van Riels". Well his observation is correct, neither have I! But I do know that when I visited the lofts and home of Jef van Riel long after the 1957 dissolution sale, what pigeons he had compared favourably in looks with the pigeons I had seen illustrated on typical Belgian-style montages. When Cois Huyskens, whom I never met, and Jef van Riel flew as partners, and through their remarkable prowess nearly smashed the great Union of Antwerp, their pigeons looked as if they were all one family. It was rumoured by many Belgians that they knew the foundation stock, yet the partners never did divulge this information. Nor did they ever declare publicly or through the medium of their two entire clearance sales what the foundation strains were. Their sales took place on two consecutive Sundays in January 1957 when the entire collection of 285 pigeons were sold, one sale taking place in Antwerp, the other in Brussels.

From the composite photographs which Jack Haylock, Dick Turner and I collected when we visited Jef van Riel's home and later the loft of his son who was also christened Jef we noted a few of his birds bore a striking resemblance to the earlier strain of the two partners that had created the strain of Huyskens-Van Riel, note the hyphen. Huyskens was a customs officer and Jef van Riel a diamond cutter. Whatever the strains used in the establishment of the strain, only they knew it, and furthermore they knew how to race them. Jack Haylock of Bishop's Stortford obtained a number of excellent pigeons of the strain, unfortunately he never lived long enough to exploit to the full their racing qualities, although in what time he had to enjoy them they most certainly excelled.

Often it is stated by writers and successful fanciers that the most successful lofts are those wherein an affinity between fancier and his or her pigeons exist. I go along with this in no uncertain terms. A successful marriage, a successful partnership are both dependent upon a liking for each other, a chemistry or to quote the "Oxford Concise" dictionary — to unite certain elements. It is the same with the most successful racing lofts. The famous Oliver Dix believed this, and proved it over the years despite the large numbers of pigeons he kept. The famous E J Spare of Wednesbury once referred to the advantages of creating an affinity between a fancier and his pigeons.

There may be a few others but relatively they are few and far between. Seldom do you notice any special emphasis on affinity, that is between fancier and pigeon. This quality is inbuilt into the environment of the most successful lofts. I suspect in some instances that a successful fancier does not even realise that the quality of affinity exists! Knowing your pigeons well, and making certain that they also know you well enough to trust you by not abusing, or upsetting the homeliness of their home is of the greatest importance. Talk to them without a hint of anger. If a particular pigeon refuses at first to do what you require of him or, her at mating time, or shows signs of nervousness, be kind to it, do not lose your temper.

Fanciers whom I have known and visited and who impressed me enormously by the manner they managed their pigeons included: Walter Hawkes of Walthamstow, who seldom kept more than 12 pairs of pigeons during a long and successful career; Bill Montague of Billericay, a turfsman; A E Sheppard of Woodford; W (Bill) Savage of Stanmore; J W Bruton of Palmers Green; Jack Slade of Enfield; Alf Baker of Wood Green; Reg Barker of Chadwell Heath, and also his father before him the famous Reg Barker of Billericay; Fred Marriott of Stechford; Lt. Col A H Osman of Wanstead; E H Lulham of Chingford; Jack Humphreys of Tottenham; Jim Biss of Brundall.

Without exception all these fanciers developed an affinity between themselves and their pigeons. Not one of them ever did speak about this phenomenon but insofar as their very presence within the loft was concerned the inmates were aware of the existence of trust, and likeness for each other in plain — affinity. The movement of each and everyone of the fanciers I have mentioned, and the manner by which they carried out their chores within the pigeon loft was never executed in a perfunctory manner. Such duties that are essential in the daily routine of the pigeon loft were carried out with the greatest possible care and consideration to the occupants. This is what is really meant by affinity or if you prefer a sympathetic understanding. The racing pigeon for me is one of the most remarkable and fascinating creatures I have ever known and I am quite sure that if you learn to recognise the importance of developing an atmosphere of trust and understanding within the loft the more successful will the partnership become, and the more satisfying will it prove for those in charge.

A fancier whom I have got to know, has become very enthusiastic about the keeping of racing pigeons. He says: "At the present time

I have a collection of pigeons of various strains or families, and am wondering whether I should now dispose of them and make a fresh start". My reply was to the effect that since he had never flown or raced pigeons, to simply try them out next year (he has a number of pigeons flying out) and learn to appreciate the routine management and training of the raceable stock. I advised him to learn all he could from a practical viewpoint, with all that you deduce from reading the many and various articles published in The Racing Pigeon and Racing Pigeon Pictorial. It is not easy at first as we of long experience should remember, we sometimes forget the difficulties we faced when we started to keep pigeons. It is even more so for those who start to keep pigeons and have no one in the family to help and advise. Today this is more than likely the case, for many of those who turn to the hobby of pigeon racing for the very first time, are, in fact, the very first member of the family to have taken such steps.

NOVEMBER

Outer Flights — Bloomfield and Fabry Logan and Old 86 — Busschaerts — Early Breeding Preparations — Show Racers

Mild weather at this time of year is helpful because with sunshine it is ideal for the moult of our feathered friends. Nevertheless one still has to be ever watchful in noting any of your birds that may be having difficulties in producing good sound silky, healthy-looking feathering particularly the quills of those highly important outer flights. But do remember this, good weather and sunshine are no guarantees if by chance any of your pigeons are found wanting constitutionally.

Perfection is the ultimate aim for the showmen so far as feather properties are concerned. Because of the work imposed upon the racers a state of feather perfection is not always possible to attain for the best looking racers. The weather plays a great part, difficult weather conditions, even changing and often unpredictable, produce severe and often punishing races. In this matter no season has been the exception to the rule of weather influence since both old as well as young bird races have provided the contestants over a wide area of the country with a number of very hard and often difficult, two-, three- and even four-day events. Consequently many pigeons return home with the "badges" (fret marks) to prove how hard they have had to work, and how much they were obliged to endure through thirst, lack of food, and the need to rest from sheer exhaustion. Often this can take many days, even several weeks to recover.

It is during this period of subsistence that the growing feathering suffers most from impoverishment through the lack of daily supplies of nourishment that a regulated process of feeding, vitamins, gritting and clean watering as provided by good management and fanciership at the home end provides. The really stout-hearted racing pigeon will not give in easily. Pigeons that are full of vitality, the hallmark of a sound constitution will strive hard to get home. Love of home is the driving force.

In the majority of cases YBs have little or no real inducement to come home at speed in the way we are able to induce or encourage our OBs to race fast home. Continual training and a succession of races, with many flying both midweek and weekends simply because they are expected to fly their hearts out and collect the loot! They do not know that we are entering them in races to bring us "owner's glory". All they know for all we can tell is that they are being constantly placed in baskets for training flights, and all too often badly designed crates and/or shaky transporters and driven to the race point as fast as the driver can without being given a real chance to have any rest whatsoever. It's enough to make them thoroughly sick of the sight of us.

Young girls and boys leave home, never to return, for some most probably because they too, like our young pigeons, consider they are not wanted! Uncomfortable homes, and uncomfortable panniers, create an environment that is lacking in a homely atmosphere of warmth, security and loving kindness. All too often the real cause of missing YBs and disappearing children! They finish up respectively stray pigeons, or reported missing sons or daughters. No doubt there are plenty of fanciers who owing to the environment they create and maintain throughout can quote often how difficult it is for them to get rid of or lose a stray pigeon. Is it because the stronger, tired, hungry and all too often exhausted pigeons — especially current year's stock — have found at last a really friendly caring environment? Maybe a faulty environment is the reason that now and again we learn of stray YBs that become consistent winners in other fancier's hands. They have found that "something" worth racing home to. Think about it for a moment.

As always I learn that many of these "missing youngsters" are now working through to their respective homes, complete with their "fret mark badges" as evidence of the extent of their travels to their long lost homes where creature comforts have been the hallmark of good management. No doubt too, the mildness of the present weather (with exceptions in some parts) will see a lot more turn up before Christmas is upon us. So long as they are healthy (and especially free from evidence of respiratory problems) with maybe only a fret mark or two, they can be given another chance. But a word of warning if you want to be doubly sure, then it would be wise to keep them separate from your team for a week or two just in case they have contracted paramyxovirus. You can still let them out, and once they are in again, return them to their quarters for feed, water, vitamins and grit. However, personally, I prefer to be safe rather than sorry and confine them, for the whole period of isolation up to a whole month.

Doubtless (as I too have learnt from experience) a number of these returnees will have learnt a great deal during their wanderings which will if given the chance prove beneficial in their future racing career. You have to assess the pigeon. By observation, you will soon be able to judge whether or not the returnee has prospered since the period of possible hunger and tiredness ended. Such pigeons that find their own way back and within a short period of time after returning quickly reveal by their very actions that they remember everything about the interior of their section, know what is their own perch and are fully prepared to assert their right to be there. Those are the ones I like to see, and welcome because I know they are glad to be back. That is one of the aspects of pigeon keeping that I thoroughly enjoy, because in the past I have remembered those that did do well for me, as well as for brother fanciers with whom I have exchanged experiences with.

Remember always no matter how well bred a pigeon may be it is absolutely imperative that it is sound. Vitality is the best word to describe soundness. Vitality is based upon a sound constitution. So important is vitality that without it you would be quite unable to vitalise succeeding generations. It matters not how much you value a pigeon, or how much its parents may be valued you quite simply have to be ruthless and eliminate any that do not measure up to a very high standard physically, as well as reflect supreme vitality.

This time of the year fanciers often are looking to make fresh purchases, always a worrying time. My advice once you have decided to take such steps is that you make absolutely certain that your latest addition is given complete isolation treatment for at least a couple of weeks. Watch them closely daily in the same way that I advised you about returnees. It matters not where the pigeon has come from. You do not want to bring into your loft a virus infection, disease, or malady. No one, including the writer, can be absolutely positive of each pigeon's health unless he or she is able, constantly and daily, to pass an examination on each pigeon. You simply have to in this day and age with ever rising evaluations placed upon pigeons commercially, carry out every precaution and safeguard against the possibility of introducing an epidemic, a virus, or bacterial infection into your loft. This is even worth thinking about, however, when you come by a new introduction. Even a stray pigeon if allowed to mix with your own pigeons can prove to be a very costly measure of irresponsibility. Leave nothing to chance. Take every possible steps to protect your stock. Paramyxovirus, paratyphoid, pigeon pox are only

Frank timing in at the lofts at London Road, Enfield.

around the corner and ready to strike given half a chance. However, I do not wish to alarm anyone, nor do I wish to be thought an alarmist, I simply stress the ever present chances that exist in order to help those with lesser experience to take precautions.

Always it seems, especially around this time of the year young fanciers and the lesser experienced appear anxious about late breds. Some have reared a few, and others have purchased bargain late breds. There are plenty about. One very young fancier wanted to know if pigeons bred so late could ever be any good as future breeders. My answer was an emphatic yes, provided they were well bred, well reared, and could be certain to be well looked after in their new home.

Some years ago I sold the remainder of the late Stan Bloomfield's pigeons, which were all of the Fabry strain. Stan only ever kept this strain. Over the years he developed a great and lasting friendship with the famous Georges Fabry of Liege and later still with Georges' now famous son Victor. Stan Bloomfield used to like to rear a few late breds. Often because of demand he would let them go, and later on regret his decision! He recalled how his good friend Georges Fabry set great store on well bred late breds, taken from his best racers and breeders. I have never forgotten Stan Bloomfield's story related again the last time he visited my home.

Stan said that in or around 1930 Georges Fabry anticipated moving from the original site of his pigeon lofts, which if I recall were situated in the attic of his pharmaceutical premises. Mr Fabry bred specially a whole round of late bred from his top breeders and racers. These late hatch babies were actually hatched late August and September 1930, and taken to the new home of the Fabrys, so well known to many visitors from the UK to this day. Despite a great deal of criticism from his closest friends, Georges Fabry, creator of the strain that bears his name to this day he had every confidence in his inbred strain and his round of 1930 bred late breds. After Georges Fabry died the loft continued under the successful control of his son Victor. In 1932 those 1930 bred late breds proved the confidence of Georges Fabry by commencing to win again top prizes and so recommenced a long and successful era for the inbred Fabry strain.

Remember too, that Georges and Victor were not afraid to try out new introductions, although as the late Stan Bloomfield impressed upon the writer, "They were very fond of inbreeding, and especially keen to obtain their own strain from lofts that have proved success with the Fabry strain".Stan Bloomfield's last visit to me and his story of the Fabry's re-establishment in 1931, or thereabouts has never been forgotten. In any event I still use the Widowhood boxes that Stan designed and had made for use at his Ware home to remind me of the value of late breds, and of inbreeding as expressed and impressed upon him by the famous Georges Fabry of Liege. Incidentally these two facets of pigeon keeping have also been of great interest to me long before I ever met Stan Bloomfield, and still do.

I make no apology for my reiteration of a statement made several times that for me there is no creature more fascinating than the present century development of the homing pigeon, the delightful bird that I of course refer to is none other than the racing pigeon. Even when I was a boy many who kept these birds referred to them as "homers". Doubtless the advent of the railways helped to bring about this change of name, from homers to racers, simply because the railways provided the first and only means of transportation to the race points. Thus the racing pigeon was born, ironically John Logan, an Irishman known to his most intimate friends as "Paddy" Logan was responsible for the laying of British railways!

Interestingly A H Osman, the first editor of the paper he founded, with the blessing of John W Logan MP and other well known fanciers named it The Racing Pigeon, and included the photo of Logan's Old 86 on the front page! The date of the first ever publication was 20 April 1898, price one penny, and Logan's Old 86 although much reduced in size, remains to this day, but not the price! The directors of the company included J W Logan MP who was chairman of the board of directors. This august group included the following famous fanciers: W H Bell (vice-chairman), R Slack, H W J Ince, F Romer and W F D Schreiber and H J Longton as company solicitor. This body were representative of fanciers from the north, the south and the Midlands. Shareholders too included a number of well known and highly successful fanciers of those times, they were A Bailey, W Bancroft, G A J Bell, E Bower, P Clutterbuck, E H Crow, G Garlick, both these last two were pigeon journalists. The former was later editor of the Homing Pigeon (now defunct), the latter named previous to the founding of The Racing Pigeon was editor of the Homing News (also defunct).

Continuing the list of shareholders were S Gibson, C Harrison, J P Higham, J T Hincks, A H Osman, W H Smith, A P Taft, T W Thorougood, J W Toft, J C Truss, P Verdon, T W Wilson, John Wones, J Wormald Jnr, G Walker and a few others whose names I cannot trace. Mr Truss was J W Logan's Parliamentary secretary at the Houses of Parliament where he spent much of his spare time "writing up" the breeding details of the East Langton loft of racing pigeons maintained at John W Logan's home which still stands to this day complete with the stables in which were maintained one of the most valuable collection in the world of hunters and carriage horses at that time.

Racing lofts were maintained over the stables, whilst the expensive Belgium imports were maintained in large aviaries of ornate iron work in the grounds. Much of Mr Truss's work on the breeding details enabled John Logan to complete his famous book Pigeon Racer's Handbook and particularly the famous last chapter "Details of the Langton Loft". Alas few of these books, retailing at 2/– (10p) each when first published exist today. This interesting book also incorporated Mr Logan's articles that he wrote from time to time, and at the request of Squills, who was then of course A H Osman. Students of pedigree and even today there are still those who derive a great deal of pleasure from such detailed pedigree information, especially of the old strains. An example of this is the John Kirkpatrick strain in particular. The work that John Logan carried out in the preparation of the few auction sales is most illuminating. Several sales were carried out in London, in Holborn, although I believe the first ever Logan auction sale was at the old Aquarium at the original Crystal Palace. However, the work carried out with meticulous care by Logan, was undoubtedly a monumental task, that pleased those fanciers who even then were enthusiastically interested in the breeding details of a good family of racers.

The final chapter in Logan's book was entitled "Details of the Langton Loft". It was a reference work of pedigrees set out in numerical order of each bird's ring number irrespective of the year of its breeding. The preparation work for this particular chapter was carried out under the instructions of Logan over the long years by Mr Truss in his spare time at the Palace of Westminster. It was equally carefully checked over by Logan as Chapter VIII in Logan's handbook confirms.

There is no doubt at all that the then editor of The RP, Colin's grandfather A H Osman, a great writer himself, encouraged Logan to republish in book form the several articles Logan had written for his friend of many years which were published in Squills Annual Diaries. The great chapter "Strains and Pedigrees', especially, is a mine of information. Fortunately the descendants of the Logan pigeons still exist to this day, although not generally so well known by the newcomers and younger members of our splendid sport. Even the great John Kirkpatrick family owed a great deal to the presence of the John Logan family in its early foundation, enhanced with Logans again in 1938, and later still with more Logan lines in 1950. There was of course, the great original Logan hen bred by the renowned Sumner Bros of Wrexham, which proved a remarkably successful producer in the foundation of the Kirkpatrick family.

My reference to A H Osman's prowess as a writer reminds me that it was in 1883 that he wrote his very first letter on the subject of racing pigeons. Unfortunately I cannot recall the contents. All I do know was that very letter was written the year my dear mother was born. A prolific writer from the time Osman wrote his first letter to the Homing News and he adopted, and it was accepted by the Homing

News editor, the nom-de-plume of Squills. Thus the year 1889 when my father was three years old, and from that time on A H Osman never failed at every opportunity to write or pen a few notes to the various papers published for the sport of pigeon racing.

In my September chapter I referred to the Litchfield race transporter crate and this reference is prompted by my re-reading once more Squills' very first Food for Novices. To think that 92 years ago A H Osman wrote, and I quote with a smile: "Don't imagine any sort of a basket is good enough to train pigeons in. I hate to see a man sending birds away in two 2½d margarine baskets. What you want is a basket as near like the club panniers in shape as possible, more particularly as regards the exit. For your ambition must be, when you put your birds in training to get them used by degrees to what has got to come later on, and if a fancier uses baskets, the lids of which lift up to let the birds fly when the birds are being liberated at the race point, they will think the correct thing is to jump upwards and bang their heads instead of running out in front".

Mr Garside of Liversidge, West Yorks writes: "Re your article on the Litchfield crates, these were before their time and should have been used extensively (perhaps we would have kept many of the race points which we have lost through the use of chippings!). We use them for training and for taking our birds to race markings. The birds are clean, their feet and rings a pleasure to handle. The bottoms on these were solid and the wire could be removed for cleaning. It is years since we met. I came down to London with the Litchfield birds when you sold them. We have a few of the old Litchfield family and these are still winning from 500 miles".

My reference to Logan's Old 86, which although greatly reduced in size still appears on the first page of the Racing Pigeon Weekly cover is and incorporated in the heading was a blue chequer cock, known as Old 86. This was not his ring number since they were not used then but his loft number. Recently, Paul Dare of Barnet, asked about the pigeon on the front of The RP. John Turnell of Cheshunt is another who has asked details of the breeding and racing of J W Logan's very earliest outstanding racer. Bred in 1879, his sire was B6, the noted Northrop Barker's famed Montauban, which won 9th prize from Montauban Belgium National and was included in the purchase of N Barker's whole loft of pigeons the same year of Barker's 9th Belgian National success. The dam of Old 86, was a Gits hen, No 646, sent by Mons Gits to J O Allen who later gave her to Mr Logan. Old 86 won 1st prize in his last four Continental races, twice 1st prize Rennes in 1884 and 1886, 310 miles, and twice 1st prize La Rochelle (444 miles) in 1884 and 1886. He also was a highly successful progenitor (forefather). There were at least 30 lines to Old 86 in the genealogical tree of Logan's 1826, the 1922 King George V Challenge Cup winner

NOVEMBER

from San Sebastian in Spain. In 1921, 1826 won 8th prize San Sebastian Grand National FC race. She was a rich red chequer hen, photographically beautiful to look at and I am convinced was designed for endurance at speed as are all outstanding long distance racing pigeons.

From the old ones to the young, it is pleasing to report that in my own circle within the sport of pigeon racing, I know personally of several young fanciers who have revealed considerable enthusiasm for the hobby. Naturally the younger fanciers are always eager and most anxious to win races. In the past, few have been known to win the very first race they competed in; the famous Vic Robinson is an example of such success. There have been others too, but just when I need to be able to illustrate such feats, as I do at this very moment of time my memory alas not being as good as it once was, has failed me, and I cannot lay my hand on my little booklet which includes such events. What a memory! Young people as most of us "oldies" know only too well have a habit of asking questions that are most difficult to answer, often impossible too! Recently I was asked, and quite seriously, by a young fancier: "How do you know when you have produced a champion?". A good question, and one of those canny ones that few if any could answer, least of all the writer. The simple answer is that no one can really tell when one has actually bred a champion until it has happened! Sounds quite daft really when put that way. The fact is that you cannot predict such an event. No matter how well you believe in your planned matings, only by trial and error, management and training with pigeons of known and proven ancestry, followed by racing will you know that a champion racing pigeon has arrived, emerged maybe would be a more suitable way in which to describe such a phenomenon.

There are a number of outstanding pigeons that have helped to make history. They have eventually been described quite rightly as champions. In early days performances in racing and training did little to encourage their owners in the belief they had at least produced a champion! One that easily springs to mind and without much effort to recall is the now famous Busschaert pigeon namely Blockbuster. Owing to the owner's decision to sell out I had the opportunity in the preparation work for Alan McIlquham's advertisement to learn a great deal more about Champion Blockbuster. The story of this very fine black chequer cock, or as I have always described him, black velvet chequer began at the London Auctions in 1979 when Ken Aldred held a draft sale of his pigeons which I had the great pleasure to conduct. Alan McIlquham attended the sale with the fixed intention of purchasing one bird from a nest pair of cocks that were bred by Georges Busschaert. These were sons of the immortal Busschaert ace breeder (also a notable racer in his day), Little Black. After carefully examining these two birds Alan decided upon the smaller of the two. Both pigeons revealed slight differences in the hand although they were nestmate brothers.

Why he chose 99545 Alan still does not know to this very day for although both pigeons were somewhat similar as befits nestmates of the same sex, his final selection was the one with several faults in handling and for type. Yet despite these physical faults there was something compelling in the bird's temperament that attracted him to Alan. After purchasing the son of the Little Black, and checking his money he decided he might have enough left to purchase a hen. Having previously decided if possible to buy one of three sisters that were penned, and thus the blue hen was selected and successfully purchased. This particular hen was considered an ideal mate for 99545. Like so many new purchases that find themselves in a new home completely shut off from the only environment and freedom they had known, furthermore closed up in a nest box for the very first time they refused to even consider each other so shocked were they with imprisonment. It was not until they were given their liberty within their new home, that they began to settle and accept their new environment.

In the first nest they produced to quote McIlquham's own words "an absolute runt". A blue pied, it left the loft on its very first outing never to be seen or heard of again! In the second nest one of the pair was a cock which was in Alan McIlquham's opinion absolutely outstanding. That particular pigeon later, but still only a young bird, became known as Blackjack, and his sire 99545, had already been given the name of Dusky. Dusky later in life escaped and was never ever heard of again. In the third nest, a pair of hens were produced, one a smokey blue the other a dark chequer and in Alan's own words "to be honest I was not in the least bit happy with the physical qualities".

As time went on and young bird racing drew closer the dark chequer cock that had been already given a name (remember Blackjack!) was one that "Mack" had great hopes for, largely because of his character. He appeared to be a great "thinking" pigeon. He was not a big pigeon but was so full of himself both inside the loft as well as outside. Unfortunately after a few weeks this great little character which had been lauded to the heights in Alan's own mind disappeared in a relatively easy race. Each and every day for many weeks after he had disappeared GB80V05468 was looked for day after day, until he became but "a far off memory". By late autumn he was resigned as being lost, or having met his fate somewhere en route. By the end of November it was doubted whether he would ever be seen again. Imagine the surprise and indeed great joy when Blackjack 80.05468, was found trying hard to find an entry into the young bird section the following July. The pigeon was to quote the owner's own words in "fine fettle". Alan McIlquham had only been involved in the sport of pigeon racing since 1978 and had never before experienced the return of a strayed "loved" one!

It was soon after Blackjack's return that Alan decided to call upon the writer with a view to obtaining a suitable Busschaert hen. Eventually I let Alan have a nest pair of hens. These were grandchildren of that ace Busschaert breeder De 45. The dam of these two hens was a super daughter of De 45 which I had by then named Golden Gift, and was bred by my friend of long-standing, Jim Biss at Hillside. Of the nest pair of hens one had been raced and showed much promise and I had already given her the name of Black Label (a favourite brand of mine), she was a very fine hen. I must say in all honesty I liked both these hens, however, there is not the slightest doubt that I did really believe Black Label to be the better hen of the two, a fact that Alan in retrospect had duly noted. Thus the foundation plan for the 1982 breeding season was laid. In the first nest from Blackjack and Black Label a black chequer cock was produced.

In addition, that season, 1982, Blackjack, then a two-year-old was put into training with a view to testing him out as a racer. Remember too that he had experienced very little racing as a young bird and none whatsoever as a yearling. Yet despite this lack of experience he was trained and raced. Blackjack flew steadily up to and including Berwick without breaking any records, yet always looked the part. At Stonehaven 378-mile event he was well fancied. He was in time to win the club event, and be in the first 15 Open of London North Road Combine. Unfortunately it was not until Blackjack had been clocked that a second rubber was noticed, having only clocked a single rubber Blackjack and his owner were thereby disqualified. The entire convoy had been re-rubbered under the Combine committee's ruling! Three weeks later he was entered for Thurso with the Combine and this 500-mile London Classic turned out a very hard race with no birds on the day. A week later Blackjack returned to the loft complete with a note attached inside the ring advising that he had landed on a fishing trawler at sea 80 miles offshore. The note included the telephone number of the trawlerman who when contacted told Alan he had picked him up on his boat whilst trawling, and that the pigeon was rescued from many gulls who had caused him considerable damage through pecking him continuously. Back on shore the fisherman placed Blackjack in a cardboard box to protect him from the gulls. He remained there for at least a couple of days between being fed and watered and was released in hope he would find his way home. Quite a performance for a distressed bird that from his appearance when he arrived back at his loft had plainly endured considerable pain and suffering.

Disease strikes

After Blackjack had been home about a week, Alan noted that he was developing sores around the eyes which gradually developed into wart-like monstrosities. Blackjack didn't appear to be any the worse for these warts. It seems that the pigeon had returned with pigeon pox, or was it gull pox!

Because Blackjack did not appear too upset Alan was more or less lulled into a sense of false security. Alas "Mack" could not have been more wrong. Within two weeks almost three-quarters of his entire stock of about 60 pigeons, of which about 30 were young birds had been infected. Many were put down. The young birds suffered the most losses. Fortunately Blackjack survived and eventually became quite distinctive as he matured with age. The black chequer cock sired by Blackjack in his first nest mentioned above was among the young birds that survived.

Although he had progressed physically he was not really all that pleasing to look at or to handle. By this time Alan decided he had far too many pigeons to handle and so took a stall at The Old Comrades Show in late November. He was offered at the princely sum of £12 but was not sold! Yet Alan had a strange feeling about this 1982 bred first son of Blackjack and can still recall to this day thinking that perhaps he had been too hasty in putting him up for sale, in fact, he felt relieved when he finally decided to withdraw him from view. On the way home Alan decided he would put the unsold youngster into the Widowhood section despite being a yearling for he had already made up his mind that he would try out this system. At one time "Mack" said of this £12 pigeon was really about the worst looking pigeon as a young bird that he had bred up to 1982!

The season of 1983 saw a transformation in the rejected black chequer cock for he flew nearly every race to and including Thurso 500 miles. He won several good prizes en route as well as winning Thurso. Furthermore he looked remarkably well when he arrived. The next day he looked terrific, said Alan. In sheer jubilation soon after he realised he had won the Thurso race, and a 500-mile race at that, he named that pigeon Blockbuster. Yes you've guessed it, the first round dark chequer cock bred from Blackjack and Black Label already mentioned several times was none other than the now famous Blockbuster. The sheer joy brought about by this comparatively modest success after getting nowhere really with the reject was to put it midly most elevating.

Although by this time he was champion in Alan's eyes he was far and away from the heights of fame and status he has attained since that memorable win for Alan. The next two seasons Blockbuster continued to bring further glory to the loft winning many prizes including sixteen 1st prizes, right through to and including Thurso.

NOVEMBER

Blockbuster also was the first pigeon ever to win London North Road Combine's Hall of Fame Award Trophy, and the very first pigeon ever to compete in the Combine that has won the coveted Double Edition Trophy twice. Dame Fortune too played her part, and I suppose too that fortune favoured the brave. It certainly did in Alan "Mack's" case! In the 1984 Berwick London North Road Combine race, the birds as is customary are marked on the Thursday. On the Tuesday preceding marking day Alan was not at all satisfied that his birds were as well as he wished. Early on the Wednesday he decided he would give them a 60 odd mile training toss and took them to Peterborough. It was far from being an ideal day for a 60 odd mile training toss.

Arriving at Peterborough by 8.00am it was raining cats and dogs. Waiting was a must, but not for too long as business appointments had to be kept, and so without waiting very long the birds were let go. Only two birds arrived on the day and these did not arrive until it had stopped raining at 9.50pm. They dropped together and they were Blockbuster and Alan's other very good pigeon The Master-Blaster.

These were the two fancied candidates for Berwick Combine and they rewarded Alan's courage by winning 8th & 13th Open Combine with 11,254 birds competing. Believe it or not Blockbuster was found in his nest box! Earlier I referred to the nest pair of hens a smokey blue and a dark chequer which were sired by Dusky the Ken Aldred bred son of the Little Black. This was purchased at an auction held in Holborn. These two were sisters, as well as being sisters to Blockbuster. The smokey blue was selected by Terry Robinson as a special case, although in all honesty "Mack" admits to this day that he was not at that time happy about either of them! Terry tried hard to buy the smokey blue hen, but was refused. That hen turned out to be the remarkably successful breeding hen known as the White Eyed Hen responsible for many winners up to and including 400 miles. She with her brother Blackjack went overseas. So in conclusion to my young friend who posed the question, "how do you know when you have produced a champion?". How do you, and the answer, one does not but with a great deal of luck they may emerge, provided they have good ancestry, usually associated with constitution and fortitude.

We are almost at the last month of the year, and for many fanciers this is the time they seem to consider it necessary to cut back on rations. I do not agree with this. Racing pigeons do not need to be fattened up like Christmas turkeys or the traditional Strasbourg goose, yet they must continue to be fed with consideration and intelligence. There are too many who consider the most severe reduction in the quantity and quality of the corn they feed their pigeons on once racing is over is best for their pigeons, when the real thought is their pocket! This is a grave mistake. Those last two primary flights especially are of the greatest importance. As I type these lines there are hundreds, nay thousands, of racing pigeons which have not yet moulted the last three flights let alone the final two. These are the highly important ninth and tenth flights.

These flights are so important as to merit the most exacting attention and the best and widest variety of mixtures. Not for me the stringent all barley, or even a wholesale depurative feed. Racing pigeons require nourishment not punishment simply because racing is over. Next year's races are won now., That is what is meant by all-year round attention. You do not have to starve your pigeons ever. This is a grave mistake. Neither is it necessary to overfeed at this time of the year when pigeons are required to be less active. For example, the reduction in daylight hours coupled with the fanciers' hours of work when for the majority it is dark when they leave for their daily employment and dark too when they return home. In any case it is the resting period for our racers. However, there are always those fanciers who have other ideas.

For instance, those fanciers who intend to mate up a few pairs in order to obtain a few early-bred young ones, and at the same time breed from the cocks they have earmarked for Widowhood racing. In event of this you would need to change the feed for the hens at least a couple of weeks before mating by providing them with a far less fattening mixture. Probably a 50% depurative mixture to which could be added a few peas and/or beans, or some of each plus a little good quality maize. Personally I do not believe it matters whether you use red plate maize, french maize, or dog-tooth maize so long as it is sound and sweet smelling. The reason for this less rich mixture is to help reduce the interior fat that has a tendency to build up, particularly at this time of year when they are less active. However, there are a number of well-balanced cheap winter mixtures available today, far more than ever before. There is therefore plenty to choose from, certainly many more than was the case 50 years ago.

Prisoner stock birds too have to be skilfully managed in preparation for the breeding period for they too have been subjected to a less active life and interior fat builds up rapidly hence the need for dieting. Such measures will help make certain that your hens do not suffer egg-laying difficulties. Under more natural conditions where fanciers do not keep prisoner stock and also do not intend to mate until the middle of February I do not consider it necessary to reduce rations until New Year's Day. However, for the showmen, and I refer to those who are out and out dedicated showmen the situation is vastly different for they have already taken the requisite steps to complete the moult by separating their prospective team of show racers as far back as early to mid-June. Consequently their best prospects are fully ready for the major show events of the coming season. Such prized possessions are fully moulted and ready to be judged. They are what is termed by the vast majority of experienced showmen as being clean moulted. These pigeons are also schooled for the pen and many of the best really have learned their trade as a result of the skilled management and patience of their owners. The top grade champions I swear know they are exceptional lookers and react accordingly!

Top showmen do not ever stint their show racers for good clean quality and variety in the feed at a crucial time of the replacement of the ninth and tenth flights, or a reduction in the normal quantity. Not at all. The showmen are fully aware of the importance of the moult and those ninth and tenth primary flights in particular. Those end flights have to be completely moulted through to their utmost natural length without showing the slightest hint of impoverishment. Sparse rations with little regard to quality and quantity would most certainly not bring forth perfectly formed outside flights. Such consequences would not produce success in the exhibition world, any more than it would in the world of racing. The successful racing pigeons whether they be short, middle or long distance, will with exception, have enjoyed a perfect moult and in particular with regard to those extremities, "the outer three" as I call them. What is more you do not have to overfeed to achieve this healthy state of affairs. The top showmen have shown by the consistency of their major show successes that they have considerable skills in management as well as excelling themselves in the art of breeding good looking, indeed beautiful pigeons. The late George Greenshields proved well enough how to win at the major shows and also to win top prizes racing often with the same pigeons. However, I am quite certain that a visit to the forthcoming Racing Pigeon Old Comrades Show will reveal to enthusiasts how skilful the show racer enthusiasts really are when it comes to producing beautiful pigeons and also their consumate skill in putting their entries down in the most superb condition.

At next year's Old Comrade Show we may for the first time ever in the United Kingdom, and possibly in the world, see a Best in Show for flown classes being awarded to a racing pigeon. It is now not so far fetched to imagine that this could happen for the modern day racing pigeon is in certain ways becoming more and more handsome. In my opinion there is no doubt of this at all. Why not? It can be done, as the famous Belgian fancier Renier Gurnay of Verviers proved in the past, and the late famous French fancier Pierre Dordin proved even more recently. Good looking racing pigeons could win in the pen as well as in National and International races. Paul Sion too was also responsible for producing a splendid family of successful racing pigeons that could also hold their own in the show pen and did, as many proved.

This year's Old Comrades Show will I am certain present stiff

198 MONTH BY MONTH — in the loft

The lofts at Enfield as they were in the late 1970s.

competition, in fact, it is possible that from the pictorial evidence seen in recent Fancy journals and what I have heard over the grapevine that it will be a hard task for the judges to decide upon the supreme champions. It does appear that fanciers today are more conscious of beauty in their racers than ever before. The pity is that the dreaded paramyxovirus is as bad as ever it was and likely to discourage the local club shows which so many racing fanciers enjoy. However, we must not be too downhearted, there is still our own great love for a good racing pigeon that is very difficult to oppose, and so, as so many have before, in times of great adversity, we will with perseverance and fortitude overcome.

The racing pigeon, and the show racer are and can be such wonderful creatures of extreme beauty and grace sufficient enough to encourage the most pessimistic to try harder to succeed in his or her efforts to produce even more beautiful, more dignified and if that were possible more graceful pigeons than ever before.

Those of us who are fortunate enough to attend the Old Comrades Show will I am quite certain see a great deal to excite them. Without any doubt it will prove more competitive than ever before. Together with specials and those wonderful trophies, plus a possible record cash amount to the Old Comrades Show top award in the flown and racing classes it is possible for a record-breaking win. Most probably it would prove a world record in prize money. At least I am hopeful for this to prove to be the case. What a great thrill for the promoters of the show, and an even greater thrill for the winning owner. It will create a new slant and a far greater interest in the showing of racing pigeons than ever before. Especially if the actual record breaker should prove to be a genuine racer, and I mean this in the nicest possible manner.

If this should happen racing fanciers more so than ever before will turn their thoughts and efforts to the production of racing pigeons that are more handsome, more beautiful than ever those that have gone before. Quite a task when one recalls as I can the many wonderful show racers that I have seen, handled and judged in days of yore, including both the Old Comrades and the People's International which alas has long since gone. As both Gurnay and Pierre Dordin proved you can produce racing pigeons with beauty as well as brains. It is however certainly not a simple task, as the famous Irish playwright once intimated to the beautiful actress Gertrude Lawrence who expressed a desire publicly that she would have liked very much to have had a child by George Bernard Shaw whose immediate reply: "but supposing the child were to have my looks and your brains"! Nonetheless, today more than ever before in the realm of the racing pigeon we do have the great strain of Dordin to assist us in such efforts. Even already there are several excellent top class show racer breeders who have proved that by crossing the most handsome specimens of this famous French strain into their established show racer families they can improve the vigour as well as the racing ability of their pigeons without losing beauty. It has been done so much that there are show racing lofts today that are more and more capable of racing well above the norm.

Long since gone are those exaggerated show types of years ago. The show racer, as well as the racing pigeon of today are much more gracefully racy in their beauty than many gone before. It is not often I can find old copies of The Racing Pigeon but during the year I came across a copy of The Racing Pigeon of October 1961. That is more than a quarter of a century ago wherein I wrote in my Month by Month of the criticism levelled by the racing fancier against the showman and the show racers of that period, fortunately too I have been able to keep the old copy in a safe place and still able to remember where it was put! Quite an achievement for me! And so for the very, very first time ever, I would like very much, with the editor's kind permission to publish in full what I wrote then simply because I really do believe it is even more appropriate than it was at that time just over 25 years ago! This is what I wrote in October 1961: "Gone are the exaggerated specimens of yesteryear . . . The pigeon for serious and successful showing must have all its parts conforming into a completely finished creature of balance and beauty, and oozing good health in a manner that cannot fail to attract the eye of the judge. No good judge can pass a pigeon that is both beautiful as well as capable of emitting a radiance of fitness, the legacy of sound constitution and sound fanciership. To win in the show pen you must plan and prepare in the same way as the successful racer over the year plans his loft routine, methods and preparation. Whether your interest lies in racing or showing never look upon it as a game of chance. It is not luck that a particular member of your club is always going to the prize table before you. That man has worked for it. To win either at the show game or the racing game, one must merit success, you cannot buy it".

DECEMBER

The 1913 Rome Race — Thorougoods
Old Timers — Rene Boizard — Minerals
Conditioning Cake — Good Diet and
Feeding — Winter Food and Water

It was a pleasant surprise to note the interest my July chapter on the Rome race of 1986 aroused for old timers concerning the famous 1913 Rome race in which United Kingdom fanciers competed. Among several letters received R Barton of Windlestone, Nr Ferryhill, Co Durham writes concerning the Prince of Rome, Mr Barton a fancier for 50 years, worked at Thorns EMI factory at Spennymore and enquired of workmates concerning Atwood Terrace, Tudhoe Colliery, Spennymore, the address of the loft site that the Prince of Rome flew home to. The house still stands, is lived in, although now modernised.

Colin Nicholson of Derby recalls: "Brook Street wherein the loft of the King of Rome stood has I am certain been redeveloped, Brook Street being part of the old West End as locals knew it". "Both the King of Rome and the Prince of Rome are exhibited in Derby Museum having been set up by a local taxidermist and contained in glass cases". This no doubt will come as a surprise to many fanciers and those of Derby and surrounding locality in particular.

Several fanciers have written concerning the breeding of the King of Rome, but each suggestion put forward seems incorrect so I have spent many hours seeking out the details. I trust too that it will interest the members of the Smithfield club who initiated this Rome race historical reference idea. Sire of the King of Rome was a blue chequer cock known as The Cup Winner. He flew Weymouth as a youngster and won a race from Swindon winning a Silver Cup hence the name Cup Winner. Sire of the Cup Winner was dark chequer cock RP05F8067 bred by Mr Driver of Burnley that was highly rated as a breeder siring a number of winners. and was also raced winning 1st Burnley East End HS, 7th Burnley FC and 9th Burnley & Dist Fed from Stafford. 8067 was a half-brother to three 1st prize-winners and a 2nd prize NE Lancs Fed winner 1905. The sire of 8067 was a blue chequer known as Pickford and carried a split ring numbered RP 1908BD647, bred by Mr Pilling of Burnley. Pickford sired winners including 1st Bournemouth, 2nd and 3rd Swindon, 3rd Jersey and 1st Nantes. The sire of 647 flew Rennes and was previously 2nd Burnley & Dist FC from Swindon.

The sire of 647 is described as Ince and Stanhope, fashionable strains at the time. The dam of 647 was bred by Mr Swindell of New Mills for whom she flew Jersey three times and Rennes. The dam of 8067 was a dark chequer hen NU99LD379, always stock and bred by P Verdon of Liverpool, from cock 195 that was sired by a 1st prize Rennes winner in the NW Lancs Fed. This Rennes winner in turn was bred and raced by J Darling of Liverpool, and bred from T W Wilson's best old blood. The dam of 195 was a hen bred by W H Bell from a well-known breeding pair numerically known as 387 and 26. 387 was bred from 1587 and 1581 by T W Wilson, whilst 26 was a sister to J W Toft's well-known 56. The dam of 379 ringed Wed 1898-1985 was from a brother to H J Longton's 1st prize La Rochelle of 1896 being from his 64 x 29 pair. Dam of 1985 was of Gits and J O Allen strains. The dam of Cup Winner was a blue chequer hen bred by Mr Hudson from two birds he purchased from the famous E E Jackson of Wheelton.

Sire of this Hudson blue chequer hen was a blue chequer RP048656 also bred by E E Jackson and was a son of a pigeon known as The Duke ringed L03DF2328, and he was bred from a brother to the famous Excelsior when he the brother was mated to a very good De Ridder breeding hen that Mr Jackson rated highly. The dam of 8656 was Lady De Wheelton, bred 1902 by E Burrows of Warrington and described as a big winner. Her main wins included 1st Wheelton, 2nd Chorley, 3rd Chorley Fed and 31st Lancashire Combine from Nantes. She was described as pure Old Clay, J O Allen, and was the old Logan blood. The actual dam of the Cup Winner was an inbred E E Jackson hen, a blue chequer ringed 1166.

Now to the dam of the King of Rome, she was a blue hen rung RP057637 bred by T W Thorougood from TWT1885 and TWT7665, these were all Thorougood's best including 61, and the famous 26a. The former first found fame by siring Mr Pulley's famous racer Elizabeth which was bred by Thorougood. The King of Rome produced well with a very good Logan breeding hen blue chequer RP11LF1550 that Mr Hudson obtained from the famous Logan enthusiast J W Shearing. Her parents were obtained by Shearing from Logan for the

DECEMBER

princely sum of £14. They were blue chequer cock 261 and dark blue chequer hen 1107 and both contained the blood of Logan's famous Old 86 and also both contained the blood of the Belgian race winner of years before Rome I. According to a letter by Mr Shearing to Charlie Hudson when he sent 1550 he stated her sire 261 resembled Old 86 his illustrious ancestor very much indeed. Both 261 and 1107 were bred at East Langton the famous home of the Logans, which is still standing and the stables too, over which many of the famous Logans were kept.

The hen 1107 actually bred winners at East Langton including 1st Alderney, 1st Granville, 1st La Roche and 81st prize Grand National. What a ridiculously low price to pay, many will think for such a splendid pair of birds, but remember that was over 70 years ago when a golden sovereign was as much as many people earned for a week's work. Those were the good old days! Furthermore not a 40-hour week or even 60 hours. The RP was only 1 penny, published Wednesday and Saturday. As one can deduce from the aforegoing details of the breeding of the King of Rome in those now seemingly far off days the accent for the majority of fanciers at that time was on distance and this was not only true for the UK, but also for the Belgian fanciers as well.

The mother of the Rome winner was 7637 bred by T W Thorougood and came from a long distance family without a doubt, She was related to Thorougood's mealy cock 7639.05 the first bird to fly over 700 miles into England day after liberation in 1908 from Mirande to Sefton, Liverpool. Liberated at 10am July 21 and was clocked in at 5.51pm July 22, velocity 835 yards per minute. Total flying time, allowing for sunset to sunrise, 9pm to 4am, hours of darkness, 24 hours and 50 minutes. The Mealy Mirande as Thorougood's 1908 21st Open National winner was ever after known really did leave his mark through his many successful descendants. Besides breeding the dam of the King of Rome, T W Thorougood bred many other good pigeons not only for himself but for many other fanciers in all parts of the United Kingdom. Initially the first pigeon to project the name of Thorougood was undoubtedly 26a, a blue chequer hen. She was without question the most important pigeon in the early foundation of the T W Thorougood strain, indeed one would be hard pressed to find a pigeon bred at The Grange, Sefton, Liverpool that did not contain the line of 26a. Although she was bred as far back as 1887 (four years after my mother was born), it is possible that even to this day descendants of 26a and that other great pigeon that one always associates with the T W Thorougoods the famous 464a, a peculiarly shaded red chequer cock can still be traced. The foundation hen 26a was bred by T W Thorougood from his No 3 cock and 226 a noteworthy J O Allen hen. The former (No 3) was bred by one of the most important early English fanciers to import from Belgium namely J O Allen. He even imported through a friend in Antwerp Gits pigeons for J W Logan, although certainly not all of them.

The No 3 cock which TWT purchased from J O Allen was a brother to Allen's famous Champion Hen, and 226 was of J O Allen's strain. Note the reference to J O Allen bred pigeons in the breeding details of the King of Rome. The renowned TWT foundation hen 26a flew a few inland races and in 1889 flew Cherbourg and then never again raced. She produced for Allen five different 500-milers. One of these flew La Rochelle, a popular 500-mile race point for Midland and Northern UK fanciers in those days, twice, and all five of these 500-milers left their mark in the T W Thorougood strain and Allens 226 was well represented through 26a in those five 500-milers, and consequently the Thorougoods and their many descendants.

Thus you will note that the line of the Thorougood's still prevails even to this day through the Biss Gold Medal strain, and of course the pigeons bred down since from the Tom Ferrar & son Thorougood family that excelled themselves in NRCC races, especially in the Shetland Isles Lerwick races. One pigeon that Jim Biss would never sell was his great favourite Old Finale, which I believe bred until his 18th or 19th year. Personally I recall handling him several times both at Wood Green Park and at Brundall, the last time when he was around the 17 year mark. He carried the old Tom Pied bloodlines and he was a genuine 100% T W Thorougood, bred by T Ferrar & son.

Jim Biss selected him as a squeaker when visiting Tom Ferrar's loft. The more recent descendants of Finale include Hillside Digit who is a grandson of Finale, and in turn Hillside Digit bred Hillside Javelin a Midland National winner for Jim Biss. Thus we span almost a century since you will recall 26a was bred in 1887.

Both the J O Allens as well as the T W Thorougoods were very important and highly influential foundation strains of the English racing pigeon. The former fancier J O Allen, together with J W Logan and Colonel Colville of Lullington being largely responsible for the pioneer work in establishing long distance racing in this country. Col Colville died in March 1886 but his widow The Hon Mrs Colville continued to maintain the loft and race with not inconsiderable success for several years after the death of her beloved husband. In the autumn of 1886 she even purchased three recommended birds at J W Logan's auction sale which took place at the old Aquarium, Westminster. Later still Mrs Colville imported a pair from Northrop Barker, helped in this by Mr Logan. No doubt surprising to many, another fancier who used the T W Thorougood strain through one pigeon was the famous J W Bruton of Palmers Green. He sent his schoolmaster friend the late Ned Carter to the home of Tom Ferrar & son and of the purchases made, I can trace the line of the

Thorougoods even to this day through a hen 54.6154 bred by Ned that he presented to me for taking him home from a club meeting one very foggy night that in the end turned out so bad I had to leave my car outside his home in Kendal Avenue, Edmonton and walk home to Enfield, about three miles in dense fog and freezing conditions. She was a grand-daughter of 51P151 bred and purchased from the Ferrars on behalf of Mr Bruton by Ned Carter.

In 1954 I sold that hen at one of my London Auction promotions published in The Racing Pigeon 20 November 1954. She had by then become through presentation the property of Ned Carter. She was a blue hen bred by W Ferrar and a pure T W Thorougood being bred from 876 the five times Lerwick cock winner of 8th Open Lerwick NRCC and four times in the money, and 1089 a daughter of the T W Thorougood pair 1331, that flew Lerwick five times winning 16th and 54th Open NRCC Lerwick; and 2201 five times Lerwick four times in the money winning 9th and 42nd Open Lerwick NRCC. Several very good pigeons were bred by me from that Carter hen. When 6154 was mated to my very good five times in the clock Thurso cock, Sunstar. She produced several very good pigeons including a very fine pigeon, rather huge in size which I aptly named Jumbo. He flew Guernsey three times as a YB, the last time in an NFC YB race from which he arrived home very badly injured in wing and legs. The following year I presented him to a friend for whom he founded a family. Each year he used to return home under his own power despite disabilities and spend the autumn and winter flying out whenever the weather allowed.

When his owner retired from the sport he was given back to me and bred several very good pigeons for me and friends. So here again, however slight, the old T W Thorougood strain still survives. the late Reg H Barker also had some of the old T W Thorougood blood through presentations and purchases of Bruton and Carter pigeons. There may still be other fanciers too who can account for the old Thorougood strain, at least I hope so. However, the earlier strains which dominated the prize lists for many years appear sparse indeed these days. At least it very much appears that way if you accept as evidence the number of sales in the Fancy Press offering sprint and middle distance pigeons for sale both private and by auction. Most offer only sprint and middle distance Belgian strains.

Only recently I received a letter on the recipe for a conditioning cake from B Walsh of Woolton, Liverpool. The cake I used to make is not now so easy to produce as it once was, simply because owing to changing fashions certain ingredients are not longer obtainable. You used to be able to purchase pea meal and maize meal, and these two ingredients appear almost impossible to purchase at least not in the Greater London area. However, you can still purchase coarse oatmeal and I have been obliged to use this instead. It is just possible that in certain areas, particularly in the farming areas, you can still purchase maize meal, someone somewhere must still grind maize! However, my own original conditioning cake comprises one pound each of pea meal, maize meal and oatmeal. Now I use three pounds of coarse oatmeal (pinhead groats) instead. To this I add half pound of glucose, half pound each of canary seed, round white millet and black rape seed. Stir this mixture well then mixing dry, add a teaspoonful of common table salt. Continuing I then add to this mixture enough chicken eggs to produce a firm mixture, usually six, if No 4 size then eight. In more recent years I have added two ounces of halibut oil. But a word of warning it is difficult to come by, and also expensive. You can usually obtain this by ordering from a good chemist. Failing halibut oil substitute cod liver oil although it is not so strong.

Place the entire mixture in a baking tin and if you are able, get your wife, mother or sweetheart to bake in a slow oven until it is firm (hard) but not burnt or it will be useless, not even fit for the wild birds for the vitamins will be completely destroyed. Store in a dry airtight tin and feed after trapping, or when feeding young or for two days preceding race. At this time of the year it is very helpful in producing a fine finish to the show birds. It will if used sparingly provide a very useful vitamin intake. More recently I have toyed with the idea of including linseed, say four ounces, and may well do this when I bake the next cake.

Old Timers
My old timer friend Fred Slade I had known from afar for many years until he retired at 84 years or was it 88 as a cart minder at Smithfield Market. Not surprisingly he knows quite a bit of history about the old time fanciers including, of course his famous brothers Jack & Harry Slade members of the late Slade Bros.

Jimmy Toogood and Jim Farrenden were others he knew so well. Fred asked of me what has happened to the old strains, where have these all gone? A good question. However, Fred is most certainly right when he insists that good pigeons have always been hard to breed no matter what the breed or strain may be! How very right he is! Another old timer in Ron Mackay of Barnet, a close friend of the late Stan Calkin, referred to the enormous numbers of new strain names which have seemingly dwarfed the old British strains. His entry into the sport of pigeon racing was around 1956 or 1957 when he obtained birds from a good friend namely Ted Prescott of Potters Bar. Those were a combination of Logan, N Barkers and H H Smith Grizzles. From those modest beginnings Ron Mackay won many Fed and Combine positions, both Western Home Counties Combine as well as London North Road Combine positions, including five 1sts in the West

Herts Fed.

René Boizard
Later still through his friendship with Stan Calkin who was probably the first to introduce into this country the French strain of René Boizard, Ron Mackay obtained several birds of this family which were bred down from such great pigeons as Louvre II, Urbino, Lionne II and Stan Calkin's own Stonehaven winner.

Since among my accumulated correspondence are letters concerning the Dordins and one in particular seeking information about the breeding lines of Louvre II. However, the information obtained from Ron Mackay has not unfortunately enabled me to produce evidence that René Boizard's four times 1st prize Barcelona winner did include in her ancestry the French strain of Pierre Dordin's fabulous dynasty. One of my correspondents who prefers that I do not mention his name says that Louvre II never won a Belgian National. Someone somewhere has, of course, got it all wrong. Since René Boizard was a Frenchman living in France he could only compete in French Nationals.

René Boizard could not, and never did claim that he won a Belgian National with Louvre II. He just could not, he lived in France! Anymore than an Englishman flying to an English loft could not compete for the Scottish National. Although in the case of the latter it is true that an Englishman has won the Scottish National but his home and loft were in Scotland when he won both a Rennes and Nantes National and that man's name believe it or not was none other than the famous John Kirkpatrick, who was born of Scottish parents on the English side of the border!

However, to return to the matter of Louvre's breeding. She was a blue hen ringed France 61.269049, her nestmate was a blue chequer ringed France 61.269050 and she won 1st prize Barcelona. Lionne II, for that was the name given to the nestmate of Louvre II, went on to win a Silver Medal at the London Olympiade. The record of Louvre II is so far as her complete racing career is concerned simply not available to me, although it is probable that had I the time I might have been able to find some further details. However, her main successes were her Barcelona perfomances. In 1964 Louvre II won 1st Club Barcelona and 5th French National, 1965 1st Club Barcelona and 9th French National, 1966 1st Club Barcelona and, 4th French National, and in her last trip to Barcelona in 1967 she won 1st Club Barcelona and 1st French National.

The dam of René Boizard's two 1st prize-winning nestmate hens was a remarkably successful racer. She was a mealy hen ringed France 55.175835 and won a Gold Ring at the Essen Olympiade. Somewhere I am almost convinced that I have read that this mealy hen, named

Ursule won many prizes racing. Ursule was from a mealy cock named Zigzag bred in 1954 and although a yearling must have been an extremely good pigeon for René Boizard to have even considered him good enough to mate to his famous hen Hetta, a blue, ringed France 49.741376 the winner of 1st prize Tolosa, Spain in 1953. The sire of Louvre II and Lionne II was a blue chequer cick ringed France 60.145714 named Homade, and was described by Boizard as a very good racing bird and was bred by a friend who stated that it was a Dordin. Homade was certainly mated to Ursule producing a remarkable nestpair of hens in Louvre II and Lionne II. According to Ron Mackay, who was a close friend of Stan Calkins. The Dordin strain was used by Stan but probably he only obtained this strain through the pigeons Calkin obtained from Joseph Zabrowski of New York.

Whether or not René Boizard, who was the chief surveyor for the town of Amiens, knew or even visited Pierre Dordin's loft I also do not know but, of this I do know that from the actual Boizard pedigrees I have seen including the one before me now brought to me through the good offices of Ron Mackay no strain name is ever mentioned by Boizard. Of this there is not the slightest doubt that René Boizard was a very clever and painstaking fancier and one who used to specialise in racing hens at the distance. And what is more with enormous success. Another of his great breeding hens was Patachou whose sire successfully flew Barcelona 590 miles four years in succession. Talisman too was a very good racing cock for Boizard and it is noted that his sire also a blue and named Audacious was bred by Stan Calkin with whom the Boizard family stayed when the Calkins lived at High Barnet.

Audacious was bred by Calkin from a pigeon he called Little Joe that won for Stan the 1964 Osman Memorial Trophy for his winning of the Combine from Stonehaven. It is possible this pigeon was named after Stan's American friend, little Joe Zabrowski, who according to the information passed on to me did import the Dordins. What is also evident is that Boizard too must have thought highly of Audacious since he mated him to Patachou the best producing daughter of René's four times Barcelona cock. Someone somewhere in the UK will doubtless be able to produce further information about the Boizard pigeons, at least I hope so.

It is quite evident for the interest shown so far as my own correspondence has revealed that the Rene Boizard pigeons have achieved a fair amount of success in the UK. Especially too it seems when crossed with the Pierre Dordin strain. I have now learned from Ron Mackay that although Homade was bred by a French fancier friend of René Boizard he (Ron) seems to recollect as best he can after so many years that the Dordin mentioned in the breeding of Louvre

Frank Hall outside the lofts at London Road, Enfield in 1959.

II and Lionne II was a Richard or maybe Robert Dordin! So there you are. Of this I am advised that the actual pedigree of Homade was not given to René. Furthermore at no time did René Boizard ever claim that his family of pigeons were Dordins.

From that great enthusiast and indeed highly successful fancier Bill Porritt of Staithes, Saltburn, I received a most comprehensive collection of literature and pedigrees, with additions of most interesting letters written on the foundation of the history of the René Boizard pigeons by René's great English friend the late Stan Calkin. In those earlier days Bill Porritt had a number of René Boizard pigeons from Mr Calkin, who in one of his letters to Bill, written in 1971, had then been a friend of Boizard for over 39 years. A long time!

However, to get back to the history of the Boizard pigeons. From the evidence before me now as I type these lines there is not the slightest doubt at all that the Boizard pigeons if not at first, were indeed very much influenced by the Dordin strain in the final years, although because of his enormous success, particularly at the distance, René Boizard considered that he was justified in looking upon his family of pigeons as the Boizard strain. Stan Calkin wrote to Bill Porritt on 25 May 1971 and I quote: "First of all I feel that it would be nice that you should know that René Boizard and I have been friends for over 39 years, and we regularly stay at each others homes, I have seen his sons and daughters grow up from babyhood until now when most of them are married with children of their own. We regularly exchange birds, and I have many sons and daughters of his many champions. Also he has sent to me many of the champions themselves, in fact, at one time he offered me the famous Louvre II but I refused to take it from him.

Of course, I have several sons and daughters of her, and in my loft today (1971) I have the cock Urbino *(whose dam was a pure Dordin bred by G Fissett of France, a blue chequer hen ringed France 62V388363, FWSH),* which was always paired to this hen until she was sold. He also sent to my loft Vedette, Champion of France for all races from 500 to 800 miles. Another hen sent to me was Lucrece, Champion of France for all races up to 550 miles. Another pair was René and Little French Lady, this pair bred me thirty 1st prizewinners, and three to win at 600 miles.

In fact, I have sons and daughters of every one of his champions, and he has many birds from my loft that I have sent to him bred from his own strain (it is the only strain I keep). Altogether I have 30 champions, or sons and daughters of his champions bred by René Boizard and sent to me. In fact, only last week I received six new youngsters and a lovely blue pied hen bred from his famous Lionne II. This strain is Dordin-based, but as René Boizard has raced and built up such an amazing long distance strain, (winner of 1st prize

DECEMBER

Barcelona almost 600 miles seven years running), 1st, 4th, 5th, 6th, 9th, 13th and 23rd French National in these seven years from Barcelona, and many other fantastic long distance winners, he feels that now he can call the strain Boizard".

Thus from this letter alone one cannot be in doubt at all of the powerful influence of the Dordin in the development and continuity of the René Boizard strain the name by which René preferred it to be known. Other fanciers who also used the Boizard strain with devastating effect racing-wise include Jan Wilcok, Frank Dell, J J D Sproul, the late Frank George, J J Perritt, Ron Mackay and George Potten who also has contributed a great deal to the interesting correspondence at present developing from my earlier letters concerning the breeding of Louvre II. In fact, George Potten reminds me that I could derive a great deal of information if I was to look up the work I put into the preparation of his sale list way back when I sold his own pigeons at the London Auction in the very late 1960s.

These pigeons included several descendants of the René Boizard pigeons. Also Woodfall Topper a Dordin, and a double 1st prize for Stan Calkin who bred and raced Topper. Another was American Laddie described as a Dordin was bred by M Disma, flew Thurso 500 miles five times winning 1st and a number of other good prizes. He contained Veddette, Lionne 30401/50, and other Dordin lines. Mention too is also made of Napoleon, bred by Lacoste of France, and Lucrece bred by Raymond of France. M Disma by the way, was a Dordin enthusiast and good friend of Boizard, Stan Calkin again was responsible for sending these pigeons to George Potten who I believe bought these in conjunction with Stan Towers.

As each letter that I have received, concerning the Boizard pigeons and there are still plenty to give further time and study to in the days ahead, I am hoping that something may emerge concerning the earlier breeding of the Pierre Dordin pigeons used by the famous Frenchman soon after the 1939-45 war when the Villa Patience Lofts were again involved with racing under the guidance of the late Pierre Dordin after his demobilisation. Unfortunately to date little has emerged other than the name of those earlier great pigeons already known to most Dordin enthusiasts. However, I will keep trying, something may turn up yet, as evidence by the enormous interest shown already of the Dordin-based Boizard strain. It is just possible that if one could tap the history of such lofts and fanciers like Raymond of Roubaix, France, also Gavdet of France, Lacoste of Bordeaux and M Disma of France, all of whom were known to be very interested in the late Pierre Dordin pigeons. Who knows maybe these columns may help to unravel the apparent lack of information surrounding the earlier breeding of the Dordin, in the same way that to date these columns have helped add to the earlier history of the Boizard pigeons whose earliest progenitors include Paul Sion pigeons. This was also a strain that latterly was used by the late Pierre Dordin through Leon, who sired the well known Neon who in turn was partly responsible in the production of Spahi, Scout, Romulus and Ramses, four top class performers in the hands of their breeder, Pierre Dordin, that have each contributed enormously in the continuity and success of the Dordin strain now so firmly established within the British Isles, as well as in France, America and Brazil.

There is not the slightest doubt at all, as evidenced by the tremendous interest shown already that those fanciers who have so selflessly gone to the trouble to contact as well as supply me with so much valuable information concerning the René Boizard strain are fully convinced of the importance that the Dordin strain played in helping Boizard to excel so brilliantly in long distance racing over a long period of time. The dam of Urbino, the mate of Louvre II until she was sold to the Louella Stud, was actually bred from a pair bred by Pierre Dordin.

Mineral Elements
Fanciers seem puzzled about this seemingly complex subject of minerals. Often I am asked do I give my birds salt at this time of the year? For my part I consider that salt is most essential. The three main mineral elements required for pigeons are, of course, common salt (sodium chloride), calcium salts and iron salts. These same elements are equally found in the various grains that we feed to our pigeons. But despite this knowledge which is elementary I always make certain that my pigeons have a wide variety of grits, at all times, and make certain of giving them one that is salty. Pigeons require a daily intake of salt. Each time they excrete they release in their urine small quantities of salt and this has to be replaced. The pigeon body-wise contains six distinct types of chemicals.

Peas, beans and tares are considered protein feed. Maybe for the novice this can be misleading. What it means is that those three grains contain a higher percentage of protein matter than does maize or wheat or dari. the heavier grains, beans, peas and tares are really the body builders. Maize, wheat, dari and barley are the carbohydrate grains. they replace lost energy units and are therefore essential for the successful maintenance of a loft of racing pigeons. Linseed is also a valuable food for racing pigeons. Linseed contains around 33% protein, and a very useful 7% or 8% fat which for the pigeon is very readily and easily assimilated. Pigeons when fed linseed for the first time do not take to it readily but I have always found they will in time.

DECEMBER

Conditioning, or trapping cake

The previous chapter which include a recipe for a vitamin conditioning cake, has created a great deal of interest. From a very enthusiastic fancier in Ireland I have even received a parcel of mazie meal! Many thanks indeed to Mr & Mrs Uprichard. From others too come letters advising where one can still purchase maize meal, but none so far concerning that seemingly invisible pea meal. From Hereford Jack Jones also advises how easy it is to obtain maize meal, health shops mostly sell it including Holland and Barrett. A friend of Mr Jones has carried out a little research on the "ins and outs" of buckwheat and in due time I hope to receive the results of this research. Buckwheat is a grain that I have used for many years. Ever since I read an article describing the physical well-being of Russian prisoners of war whose greatcoats were lined in the heavy seams with buckwheat! Yes, really! The article appeared in The Telegraph apparently and according to this article in which it was written that one of our own generals noted that the Russian prisoners with much self-discipline in serious times of food shortages allowed themselves a small ration of the grain to offset the lack of fresh food and the consequence shortage of vitamins. However, unless a particular conditioner is used, in which buckwheat is included, it is still not used by fanciers as much as it could be, like linseed it is all too seriously underated. Despite what some may think I really believe buckwheat has a positive value so far as racing pigeons are concerned. But they (the pigeons) have to be educated into accepting this curiously shaped cereal.

Thousands, probably millions of words have been written and published in the Fancy Press, both weeklies as well as the monthlies concerning the corn fed to racing pigeons. We also have the many and varied articles written proudly by the successful fanciers and contained in the annual Squills Year Book. The ever intriguing and vexing methods of feeding in the majority of cases are really not so well detailed. Invariably it is fancier's likes and dislikes of certain kinds of grain they usually refer to. Although as many of us know by a study of these articles and a perusal of the corn merchant's sales lists and advertisements that there really are many different preferences shown by fanciers. Few really lay down in concrete form what really constitutes a suitable diet for the feeding of racing pigeons. The many different types of mixtures made up by the corn merchants is largely due to the demands made by the fanciers and the merchants obvious desire to earn a living or meet their labour costs by meeting the many different demands made upon them. The ever varying types of mixes is almost as diverse as the many and various articles devoted mainly in gardening books for amateur gardeners in order for many to be able to produce high quality prize-winning garden produce. But these articles do try to explain the reasons for the advice meted out. Such explanations are seldom if ever included in the articles written by pigeon fanciers.

This is rather a pity and an oversight that I hope the successful will consider whenever they are invited to contribute articles to the Fancy Press. If you have a collection of pigeon books, or perhaps a year's issues of Racing Pigeon Pictorial you could be kept busy for several hours, even days, writing or noting down the many and varied types of corn favoured by the successful. Some even mention the mixes they use. Whether or not they feed by hand or by hopper is sometimes stated. Also if they feed once a day, or twice a day. Or as with the system of hopper feeding they allow their birds to have access to feed all day long. Yet quite seriously not too many divulge or explain in detail what they really do. Seldom do any offer a positive explanation of why they favour certain cereals, or particular proteins. Seldom if ever do they divulge how they feed their pigeons in the detailed manner that so many of the lesser successful, and especially novices would like really to learn about.

Some no doubt do this purposely. Others, and I believe the majority do not even consider it important enough. My own view is that quite often they have not the patience to put it down in writing. Many there are who will still refuse to include maize in their feed. Others are adverse to the inclusion of wheat in their mixtures. Whilst others do not use beans whatsoever. Whilst some feed only beans a little maize with a few small seeds from time to time. Again many refuse to use small seeds at all. Tares are probably less popular than they once were, largely owing I believe to price and the lack of quality as opposed to the old giant Goa tares grown in Sweden but no longer exported to the UK. But as I have stated few of the successes explain in detail their likes and dislikes, or the detailed applications of their methods.

Yet for the many who reach out despairingly for such advice and information it would prove a great blessing. Circumstances are so different for each and almost everyone that someone somewhere would be able, if fanciers heeded my appeal to discover a system that should meet their needs and circumstances for their hobby admirably. However, this state of affairs it not confined to the world of pigeon racing. One does not often see published a system of feeding and management meted out to their valuable charges by the most successful racehorse trainers. There is probably a good enough reason for such reticence since it is their livelihood. Or for that matter greyhound trainers who seem somewhat loath to impart knowledge of feed and training their charges. At least that has been noticeable to me. I certainly cannot recall any detailed system of training published in the sporting pages. Maybe if you read the works of various sports writers who specialise with racehorses, as I do whenever possible, you will come across a line considered of news value included by a keenly observant reporter wherein a mention is made that a certain trainer practises a particular method or a revoluntionary form of training his horses. Even to the mention of specially treated oats laced with stout and fresh eggs. Or maybe a linseed mash that has been laced with grated carrots. And of course many no doubt have read of the odd character in a racing stable, a Red Rum type, that enjoys his pint of stout!

Turning to greyhound trainers who have established reputations through their outstanding classic performances on a par with the late Stan Biss Snr. trainer of greyhounds extraordinaire, or the famous Leslie Reynolds, who were reputed by reporters to give their most outstanding charges a couple of pounds of top quality rump steak that had been previously soaked in a good quality sherry for two days in succession prior to the big event! Others too have been reputed to give their top dogs a pint or two of stout! Likewise as one can expect you hear of, and sometimes read about pigeon fanciers who proclaim generously to reporters that they feed their top racers with canary seed that has been previously soaked in sherry. Others too that make it a practice to feed hempseed that has been thoroughly soaked in glucose water 12 hours before use, and then before being fed is dried off with powdered yeast. From experience not such a bad idea for sprinters and middle distance pigeons either. But such tips, styled by some as "secrets" are of little use to the novice or lesser experienced simply because they have not reached the stage in their development as practical fanciers whereby they are able to recognise condition and the possibility of "form", that magic word that spells the ultimate in a good pigeon. That probably sums it up — the importance of a continuing policy of intelligent observation.

The production of form in a racing pigeon is probably one of the most difficult subjects to describe on paper, especially well enough for the novice, newcomer, and the unsuccessful to understand fully. However, I will try again by explaining although making a contrast so strange that it may appear ridiculous to some, yet nevertheless in the name of dedication to a sport, or better still a hobby, that has given me a great deal of pleasure for these past 70 years, a comparative interpretation that I trust you will understand. Try to imagine the following. Take a piece of hardwood that you wish to smooth into a special shape. You can either use a smoothing plane, or coarse sandpaper, followed by a fine sandpaper.

Better still buy an electric smoqthing machine. Hard work indeed, but the will to succeed coupled with the right tool and sufficient energy will produce the shape and finish you require for as near as possible perfection. Now imagine your pigeon, or pigeons, that are your selections for the forthcoming season's races. At present for many they are like a piece of softwood. Without muscle or firmness associated with racing condition. Flabby if you like through good autumnal feeding, and lack of exercise. For a pigeon to be able to sustain flight at speed without stopping through tiredness, it has to be thoroughly fit and muscular. The comparison may sound silly, but wait for it, and think about it! A pigeon that is fit for racing will feel, in hand, tightly feathered and hard. To the eye it will appear sleek, and scintilatingly brilliant with snow-white cere and wattle — in fact, a joy to behold. A pigeon that is muscularly fit can fly with ease in a naturalness that personifies fitness and power the result of co-ordination of mind and physique to enable it to fly strongly even against the wind. A truly fit racing pigeon will appear sleek smooth, yet feel hard in the hand. To quote my favourite expression: "Hard as a brand new tennis ball". Almost slimline would perhaps be a better description of a fit pigeon. Streamlined probably suits even better still. In bringing about this transformation you are creating a subject that appropriately has become aerodynamically suited to the task of racing without the greater effort that a less streamlined weightier pigeon would prove itself to be. Such a transformation can only be achieved through the reduction of fatty tissues and a conversion of such tissues into muscle. Only by regular bouts of exercise and judicious training can the muscles reach a state of hardness that one associates with the physical fitness of a class athlete who has been trained to perfection. Only work and sufficient suitable food will achieve this. It goes without saying, that the loft or pen and the cleanliness of the housing must be warm and dry, yet ventilated sufficiently to provide a healthy environment. For those who favour the Natural system the ability to race and win prizes will be further enhanced by your knowledge of the various conditions you have noted that appear to suit certain pigeons best.

As I have already stressed only a policy of patient and indeed painstaking obverservation of a consistently perservering nature will enable you to accumulate such information. This is what is meant by good fanciership. As I have also already written, and so many times in the past you really do have to be dedicated; this requires time and tremendous determination. Those so called "secrets" or useful tips that have been carried out by confident fanciers who know how to "get one ready" are handy to know! To suggest such ideas to a novice or even an unsuccessful long-term fancier who has not yet learned the art of feeding and is not able to recognise condition would be both foolish and wasteful. When you have learned through experience of feeding and exercise to appreciate what is meant by condition and form, then is the time to consider the implementation of such tips and ideas. Fanciers, especially the unsuccessful believing that the

DECEMBER

successful possess secrets are wont to seek such advice when they would serve themselves better by experimenting in order to learn to recognise the appearance of condition brought about through a combination of training and feeding. Observe all you can, note what you see, and continue to record the results of your observations until you get it right. You have to produce muscular fitness, and you can only achieve this through regular bouts of exercise, followed by gradually increasing bouts of training. This spells work for your pigeons, not to overlook your own efforts. But it is important to remember you cannot expect your charges to work hard for you if they are hungry, or flying on reduced rations. This is the mistake that many make. Too many in fact. They overwork their pigeons, and unfortunately underfeed them. This is just as fatal as overfeeding without the required amount of work rate.

It is very difficult to get this over to the uninitiated, or the comparative newcomer. You will not produce form within your team unless the optimum conditions prevail in your establishment, or loft. I will conclude with my own observations made over the long years. Ideally I believe it is best to give your pigeons at least two meals a day, especially in the long winter days. Water is often a cause for alarm during the winter so if you prefer to hand feed them a light breakfast when you attend to frozen fountains it will be appreciated by your pigeons. It is comforting as well as kindly. The final feed can be given as late as possible, and of course according to your working day. If on the other hand you enjoy the luxury of employing labour, and have installed an electrically controlled water-heated system to prevent your water fountain from icing up you have no worries. But very few are in this position. If as so many do you spend long working hours away from the loft you will have either to obtain the help of your wife or think about the cafeteria method. The hopper system, therefore, makes sense provided you have hoppers that will allow the mixtures to trickle slowly into the base of the hopper. The Eltex automatic metal hoppers are quite useful for this provided you take steps to keep your wing nuts free to turn with a little oil now and again. This will enable you to adjust the flow of corn. Or better still apply a little tallow to the wing nuts and fixed bolts. This I have found to be best in the war against rusty nuts and bolts. You can also design wooden hoppers that will allow a trickle of corn to drop into the bottom and thereby prevent the birds from scattering corn that can so easily become polluted when searching for their favoured grain. Make certain that the bottom of the opening is two-and-a-half inches from the actual bottom of the hopper. Although there is much to be said for hopper feeding you do have to reckon with vermin, especially mice. These problems you must work out for yourselves.

The ideal would be to design a hopper that can be fully closed to vermin during the night. Equally the ideal hopper is one that will enable the birds to eat without gorging themselves. The famous H Seaton who specialised in the breeding of Oriental Frills, and who lived at Woodford Green, Essex described the benefits of hopper feeding in a book he wrote in 1912 entitled "Oriental Frills". Appended below is what Mr Seaton wrote: "A hopper that will equally serve the purpose of a more elaborate one can be made from a Hudson's soap box, by merely sawing three slits, one at each end and one in front two inches wide, from the top to within two-and-a-half inches of the bottom. Put the food in the box, shut the lid, and the hopper is complete, and will hold a week's supply for a dozen pairs of birds". You must therefore work out more or less the size you will require according to the number of birds you keep. It is a long time since I saw a Hudson's soap box but to the best of my recollection they were about 14 inches long by about ten0 inches wide and ten inches deep. Not enough thought is given to the design of hoppers these days but I am certain there is a market for the ideal design. For those fanciers who do decide to use a hopper that is automatic make certain the corn is not able to flow too freely otherwise you will have wasted and polluted corn scattered over a wide area and become soiled with excretions. This is both wasteful as well as dangerous to the health of your flock. Unlike feeding by hand, your stock will definitely not gorge themselves through long waiting hours between visits to the loft. Hopper fed birds without question eat far less than those fed by hand, unless you go out of your way to keep them hungry, or at least stint them unnecessarily. Remember too that automatic hoppers are ideal during the breeding period. With hoppers that are designed for automatic replenishment as each grain eaten is replaced, the birds will invariably eat a little in the morning, and again mid-day, with a fairish feed at night. There is not the slightest doubt that pigeons in the wild feed more than once a day.

A pigeon loft does not have to be palatial. Many lofts up and down the country bear testimony to the amazing ability of the racing pigeon, in being able to find its way to its loft no matter how well hidden, or camouflaged it may be.

As I type these lines I have in mind George Swann of Bush Hill Park, whose pigeon loft is almost hidden from view. It is completely screened off by a vigorous evergreen hedge, yet as in other years past George has had the distinction of getting his full team of six (or maybe it was seven), home from Pau, 570 miles in 1989.

George is like his pigeons, tough. A career soldier he saw service in India, the Middle East, the Western Front, and a long four years as a prisoner of war in Poland. Since his return to civvy street he becane a first rate bricklayer. Now almost 80 years of age he still recalls vividly his war-time experience and his long march from

Eastern Poland. George also recalls well the pigeons he managed to catch and keep in various parts during his regimental days. The good pigeons, or to be more precise, the best ones, are the tough ones just like George, and his own beloved pigeons.

Success in pigeon racing depends on a combination of good birds, suitably fed, that are well trained and properly housed. Not that George's loft is much to shout about but it is dry and well aired. Probably housing is the most important subject connected with the management of the racing pigeon. The finest bred pigeons in the world will soon deteriorate if they are maintained in an atmosphere of stuffiness. A stuffy obnoxious smelling loft is probably the greatest deterent, or handicap, in the obtainment of success in the successful racing of our feathered friends than any other. A loft has to be really dry and well ventilated. George Swann's loft in Bush Hill Park is both dry and well ventilated! What it lacks in splendour it gains in practicality. George's loft has always been bone dry, and full of holes — in plain it is full of oxygen, owing to the ventilating cracks!

If your loft is well ventilated, and of course, it goes without saying, that our pigeons are of good ancestry, success follows. Furthermore, it will continue to follow over the years. Today with pigeons costing many hundreds, even thousands of pounds, it naturally follows that many attractive and most eloquent worded advertisements are bound to be found almost weekly announcing magic pills, potions and fabulous elixirs.

Blood purifiers abound in all sorts of guise. The elixir of all is actually free, and thank goodness is non VATable. It is known as *OXYGEN*. It is a wonder that someone has not already thought about bottling it up and given it a special name and offered it to the Fancy for pounds and pounds!

Oxygen is the finest elixir you can give to your pigeons. Lofts that are badly designed, and many are, will not bring you success. Lofts, pens, or whatever name they are given in various parts, that are without sufficient ventilation, create overheated, stuffy, conditions within, especially so when they become grossly overcrowded. Think about it. Oxygen comes from fresh air. Without it the loft becomes stuffy and smelly. More colds are caught by the human race through overheated and crowded halls, homes and living rooms, than are caught in the great outdoors.

Imagine what contaminated air will do to your own health. Try to imagine also what it will do for your pigeons, especially in the autumn and winter when your birds are perforce obliged to remain confined. Breathing contaminated or foul air all day and all night when they should be breathing in pure oxygen. Far too much damage is done to racing pigeons in the winter months than many imagine. There are too many fanciers who really do believe that you must keep racing pigeons warm in the winter. Comfortable yes, but not overheated like a greenhouse that has to house valuable orchids, or free from frost pelargoniums, or rare hot-house specimens. It is important to remember that pigeons produce a tremendous amount of heat when confined. Try to imagine what it can be like with each one radiating 107°F which is the temperature of their bodies. If this does not cause you to reflect, and maybe think more about the well-being of your pigeons then think about this. Try to imagine what your own house would be like if you had 100 heaters all burning throughout the night at a regulated heat of 107°F. For that is what you would be doing if your loft is badly ventilated and you had 100 pigeons packed into a loft that was too small.

In plainer terms an overcrowded loft without ventilation would be a place without enough oxygen and this is dangerous to the well-being of your pigeons for next year's races!It is a most un-natural way for racing pigeons to spend their nights under such conditions. Remember that nature has endowed them with a natural insulation against the cold. Pigeons that are housed in ideal conditions will prove this by producing superior feathering. The feathers are an intricate covering that will protect them when the temperature drops in the night. The feathers allow for temperature changes. When you go out during the cold weather you automatically put on extra clothing, and when you get back in the cosy warmth of your home you remove your overcoat. Now a pigeon cannot do this.

Oxygen is used by the pigeon to purify the blood which is passing through them continuously. To be able to live healthily a racing pigeon through the constancy of oxygen which is collected from the air is able to purify the blood. Oxygen is the most valuable elixir of all. Remember without it the blood cannot be purified.

Many of those weak squabs that are an annual source of complaints, and all too frequently reported, result from by loss of vitality in the breeders. In my opinion this is largely owing to pigeons, stock pigeons in particular, that have been subjected to badly ventilated homes (lofts or pens) during the long winter months. Dead in shell are more often than not caused by parents that lack vitality. There is a great deal more that could be written on the subject of fresh air for racing pigeons, but time, as ever, is against me, but I do hope fanciers will give the matter much thought.

One of the greatest problems today is that all too many fanciers consider that the Continental fanciers are mainly concerned with Widowhood racing (cocks only) and that their lofts (from photographs) appear to be without ventilation. This is far from true. They may not appear to be ventilated, but they really are. The ventilation system is regulated by louvres and false ceilings within the roof structure. Furthermore, as I have noted when visiting both Belgian and French fanciers' lofts, alas far too many years ago, I noted how few pigeons are kept in each section. Many of the lofts I have visited have several sections, but few pigeons in each.

As I mentioned earlier a pigeon loft does not have to be palatial. It simply must respect your neighbours who should not be subjected to the constant sight of a building that is dilapidated. It matters not how simple the design of a loft may be so long as it is practical.All lofts in my opinion should incorporate a look-out pen, and the facilities of helping your babies to develop "spot location". Far too many babies are lost because they are unable to see outside the loft until they are released from the loft for the very first time and this is most dangerous. More of this when I can get round to it.

Remember to keep your numbers down to a minimum. Quality is what you want, not quantity. Tough racing pigeons are capable of flying successfully in head-winds, and not dependent entirely upon blow homes, there are far too many of this sort already!Often and especially in the next month or two when many pigeons will be sold or change hands, including many old pigeons, one is asked such questions as "How long will a proven racer continue to fill its eggs?", or "How can you tell if an old pigeon is capable of filling his eggs"? It really depends upon the pigeon. Some will fail to fill their eggs once they have reached eight, nine or ten years. Many never get the chance to prove their fertility one way or the other. Doubtless, many of these would not be worth breeding from anyway! Some old cocks have really made history by continuing to turn out good pigeons year after year. Among the several who have sought advice on this subject is one young fancier who has been given a very good racer that has flown exceptionally well, but is now entering its tenth year, If the donor or previous owner assures a young fancier that it has never failed to fill its eggs you can in the majority of cases be certain the truth has been given.

Nonetheless, despite the credit side, there is still the mental problem. Will the move to a new home, together with its changes and the differences of climate, water and lack of old familiar scenes, people and sounds especially voices, not also to overlook a completely different aspect prove too much? An ageing process too that accelerates daily may prove enough to prevent any further chances of fertility.

Personally I can lay claim to considerable experience with the keeping of old pigeons, hens as well as cocks. You can never really tell when you may discover another Hyperion, which continued to sire winning horses up until his 29th year. The descendants of Hyperion can be found in many different countries all over the world, still winning races despite the span of 54 years since Hyperion first became a fashionable stud stallion for his proud owner, Lord Derby. Hyperion was put to sleep in 1961, but his descendants continue to figure in the breeding of many outstanding racers of the present time — so powerful are the genes which account for health, constitution and temperament, which are the qualities that combine to produce champions, be it a thoroughbred racehorse, a racing greyhound or a racing pigeon.

Old pigeons, provided they are well bred, well proven, and of a sound mind and constitution are more than capable of reproducing quality racers, and most certainly worth preserving with if you have a great deal of patience. You can never really tell when you may produce another Old K42, my old Logan cock that was bred for me by my old friend Alfred Hancock who was a great admirer and enthusiastic breeder of the early John W Logan pigeons. Old K42 hatched April 1943, bred vigorous pigeons that flew and won good races until his 18th year. Even then he did not die, but met with an awful accident by flying into a TV aerial. Recently three descendants of Old K42 were three of only six pigeons recorded in a Slough SR club on the day from Sennen Cove, flying 246 miles to win 1st, 3rd & 5th Club for an old friend Bert Sweetzer who bred these three pigeons from his Hallmark family. All three trace back to my old favourite, great racer and breeder, Old K42.

John Lynch of Enfield, who together with his lovely wife Sarah, and son Robert, recorded a good win with yet another descendant of Old K42. These little snippets of news please me enormously. Young Brett Chapman is yet another, who, in 1987 as an 11 year old, won £312 in the North London Fed Futurity race, as a result of buying a descendant of Old K42 which I presented to the Fed's Futurity Auction Sale. This was a breeder/buyer sale. The result of this year's event, again from Morpeth, as it was in 1987 flown in conjunction with London NR Combine, is awaited with great interest, especially by the writer. Hopefully too, Pat Newell, one time successful Wood Green flier now living in Essex, phoned to advise that one of the youngsters I presented to the 1988 North London Fed breeder/buyer sale that he purchased was clocked on the day from Morpeth.

Another great pigeon that springs to mind is John Taylor's old blue cock Faithful, which sired John's famous double Barcelona winner, Champion Newslad. Selbourne Combellack of Falmouth purchased Faithful, then I believe around 13 or 14 years old. The wisdom of doing so may not have appealed to many by virtue of the price paid, but purchase him the Cornishman did, and fortunately he filled his eggs until he died at 20 years of age. The change of loft and surroundings did not upset Faithful one little bit.

Previous to the sale of John Taylor's birds at the London Auctions, Selbourne Combellack had already made a few purchases from John Taylor direct. Once these birds began to acclimatise themselves they began to assert their influence in no uncertain terms. Another shrewd move by the quiet Cornishman, whom I got to know quite well as a result of his several visits to the London Auctions in search of John Taylor bred pigeons, was proven later after he purchased Thurso Queen from Jimmy Treen who had announced his intention to retire. Thurso Queen flew the always difficult race point Thurso into Cornwall no less than three times. She flew Thurso 1959, 1960 and 1961 winning in her last effort 1st Club, 1st Fed only bird on second day, when only three birds made it in race time. With the progeny of several others from the loft of Jimmy Treen, and the purchases from John Taylor both privately and at auction Selbourne Combellack made quite a name for himself. In the last ever conversation I ever had with Selbourne he was keen to emphasise the great value he placed upon his purchase of Faithful then at that time around 16 or 17 years.

I have tried hard to locate my copy of John Taylor's London Auction sale list, or I would be more certain of the age of Faithful when I last spoke to Selbourne. The real reason for reference to Selbourne Combellack is to illustrate that an old pigeon, especially a cock can prove a sound enough acquisition provided the bird in question is soundly constituted. Faithful impressed the Falmouth fancier so much that in one of his rare conversations with me he stated that he had at theat time, around 1966/67, collected around 30 pigeons bred by John Taylor. But in all fairness the purchase of Jimmy Treen's good Thurso hen was also a very shrewd buy. When the majority of Cornishmen turned their birds from the North Road to the East-West route the John Taylor-cum-Jimmy Treen Thurso-Queen based family did not disgrace themselves by any stretch of imagination. The Meudon Lofts were consistently sensational at the distance and especially when the race was a difficult one. Before turning to the East-West route Selbourne Combellack had already made his presence felt at the longer race points including Portsoy which the Falmouth fanciers included at one time in addition to the Thurso (590 miles) race. The Portsoy race flown in 1967 or 1968 Selbourne won with a pigeon he named Sky Rocket, a light blue chequer hen. It won the very hard Portsoy event, a distance of 530 odd miles after more than 16 hours on the wing, only two birds home on day in the Fed, and not more than a handful in race time. Those were the kind of performances Selbourne Combellack used to return with his Champion Newslad family crossed with the Jimmy Treen Thurso Queen almost 600-mile winning blood. Not bad when you consider how much faith and trust Selbourne had in Old Faithful, the sire of Newslad. Remember too that Faithful was bred in 1946 ringed NURP46CC7751 and changed his home when almost 14 year of age being bought by Selbourne Combellack on 21 November 1959. (After two hours searching I have just found my copy of The Racing Pigeon dated 14 November 1959).

Champion Newslad was the very first pigeon to be sold for £500 at auction in the United Kingdom. A week or so after the Belgians, through the rate of exchange evaluation of the pound, claimed the record with the sale of a pigeon for £501. Things internationally have changed a great deal since those seemingly far off days. When you consider the price Champion Newslad brought for a double Barcelona 797-mile winner including 40th Open International Barcelona, I cannot now imagine the Belgians coming back this time with a higher claim of a purchase price for a Belgian pigeon in the same way they did after I sold Champion Newslad for the famed Barcelona fancier John Taylor of Walsall for £500. A great deal can happen in 30 years.

Returning to the question of old pigeons Faithful was bred from a three year old cock 43.7571 and a six year old hen 38.9089 bred by the famous J L Baker from 1019, a prize-winner in Midland Counties Combine four years in succession. 1019 was mated to J L Baker's Mirande hen. 1019 was a grandson of 99, the 12th Open National San Sebastian winner. The sire of John Taylor's Faithful (7571) was bred by Jenkins & Cooper of Tipton, whom I never knew or met but they established a good family on birds from the lofts of W Simms of Prestfield and Dan Onions who both in turn obtained their birds from J Platts who owned early N Barkers and Van Cutsems. The N Barkers originated from Longton of Denton.